Technology Choices

Acting with Technology
Bonnie Nardi, Victor Kaptelinin, and Kirsten Foot, editors

Tracing Genres through Organizations: A Sociocultural Approach to Information Design, Clay Spinuzzi, 2003

Activity-Centered Design: An Ecological Approach to Designing Smart Tools and Usable Systems, Geri Gay and Helene Hembrooke, 2004

The Semiotic Engineering of Human Computer Interaction, Clarisse Sieckenius de Souza, 2005

Group Cognition: Computer Support for Building Collaborative Knowledge, Gerry Stahl, 2006

Acting with Technology: Activity Theory and Interaction Design, Victor Kaptelinin and Bonnie A. Nardi, 2006

Web Campaigning, Kirsten A. Foot and Steven M. Schneider, 2006

Scientific Collaboration on the Internet, Gary M. Olson, Ann Zimmerman, and Nathan Bos, editors, 2008

Acting with Technology: Activity Theory and Interaction Design, Victor Kaptelinin and Bonnie A. Nardi, 2009

Digitally Enabled Social Change: Online and Offline Activism in the Age of the Internet, Jennifer Earl and Katrina Kimport, 2011

Invisible Users: Youth in the Internet Cafés of Urban Ghana, Jenna Burrell, 2012

Venture Labor: Work and the Burden of Risk in Innovative Industries, Gina Neff, 2012

Car Crashes without Cars: Lessons about Simulation Technology and Organizational Change from Automotive Design, Paul M. Leonardi, 2012

Coding Places: Software Practice in a South American City, Yuri Takhteyev, 2012

Technology Choices: Why Occupations Differ in Their Embrace of New Technology, Diane E. Bailey and Paul M. Leonardi, 2015

Technology Choices

Why Occupations Differ in Their Embrace of New Technology

Diane E. Bailey and Paul M. Leonardi

The MIT Press
Cambridge, Massachusetts
London, England

MIT Press books may be purchased at special quantity discounts for business or sales promotional use. For information, please email special_sales@mitpress.mit.edu.

This book was set in ITC Stone Serif Std by Toppan Best-set Premedia Limited, Hong Kong. Printed and bound in the United States of America.

Library of Congress Cataloging-in-Publication Data

Bailey, Diane E., 1961–
Technology choices : why occupations differ in their embrace of new technology / Diane E. Bailey and Paul M. Leonardi.
pages cm. — (Acting with technology)
Includes bibliographical references and index.
ISBN 978-0-262-02842-4 (hardcover : alk. paper)
1. Technology—Public opinion. 2. Technological innovations—Social aspects.
3. Vocational interests. 4. Organizational behavior. 5. Automation. I. Leonardi, Paul M., 1979– II. Title.
T14.5.B287 2015
331.25'6—dc23
2014021818

10 9 8 7 6 5 4 3 2 1

Contents

Acknowledgments

Our first thanks, and our deepest thanks, go to the engineers and engineering managers who agreed to participate in our study, who tolerated us with some amusement as we sat at their elbows and watched them work, who graciously answered our many questions, and who, ultimately, taught us quite a bit about engineering work and technologies. Without them there would be no study.

Our next thanks surely go to Steve Barley, our dear friend and collaborator on so much of this work. Many of the ideas in this book were ideas that emerged in our joint conversations in our offices, in our homes, and over the phone. We could ask for no better colleague in the research process: From details of field notes and sites to struggles with theoretical framing, Steve played a huge role in what we did, what we learned, and how we told our tales.

Our research team was superb. Among students on the team, Fabrizio Ferraro, Julie Gainsburg, and Menahem Gefen observed structural engineers; Mahesh Bhatia, Jan Chong, Carlos Rodriguez-Lluesma, and Lesley Sept observed hardware engineers; and Vishal Arya, Will Barley, Daisy Chung, Aamir Farroq, and Alex Gurevich observed automotive engineers. Daisy Chung and Jeffrey Treem interviewed automotive engineers. Additionally, eleven R&D personnel from IAC collected data via field observations at the firm's Michigan headquarters. Jan Benson, John Caféo, Ching-Shan Cheng, Mike Johnson, Bill Jordan, Hallie Kintner, Mark Neale, Susan Owen, Dan Reaume, R. Jean Ruth, and Randy Urbance participated in this effort. Hallie Kintner spent a semester in residence at Stanford to work with us on early data analysis; she also interviewed engineers in Sweden. We thank everyone who worked on this project so diligently, and we particularly appreciate that they maintained their good humor, high energy, and curiosity throughout.

This project had its genesis in the Center for Work, Technology and Organization at Stanford University, where Steve and Diane were on the faculty and Paul was a student at the time. We want to thank our many colleagues there, especially Pam Hinds and Bob Sutton, for their enthusiasm and support, and for making Stanford's WTO a wonderful place to be. Frequent WTO visitors Amy Edmondson, John King, Gideon Kunda, Wanda Orlikowski, and JoAnne Yates, and nearby colleagues Ray Levitt, Sheri Sheppard, and Bruce Wooley, were strong supporters of this research; we thank them sincerely. Bonnie Nardi has long been appreciative of this work; she, Victor Kaptelinin, and Kirsten Foot, the editors of the Acting with Technology series, and Margy Avery and Katie Persons at the MIT Press guided us expertly to make our book happen. We thank them for their embrace of this project.

Diane would like to thank Andrew Dillon and Bill Aspray for making the University of Texas at Austin's School of Information a superb locale for writing this book and exploring new interests. She thanks her parents, Bob and Molly, whose work stories told at the dinner table prompted her interest in what people do for a living and how they do it. Diane thanks Anika for sharing her high spirits. Without her husband, Sushil, who lent a ready ear to stories of this project's triumphs and woes and whose care was unfaltering, Diane never would have completed this work. Paul would like to thank colleagues Noshir Contractor and Michelle Shumate for discussing various aspects of the book, as well as Barbara O'Keefe for her generous support and encouragement. Rodda, Amelia, Norah, and Eliza (who arrived just as we were finishing this book) were also constant sources of inspiration. Paul owes them all for making his life so wonderful.

This research was made possible by several grants. We thank Suzi Iacono and Bill Bainbridge at the National Science Foundation for their support of this work. The NSF provided funding via grants SBE-0939858, SES-1057148, IIS-0427173, and IIS-0070468. General Motors provided additional funding through the Stanford-General Motors Collaboratory.

Portions of chapters 4 and 5 draw upon earlier versions of our work that were previously published, and we gratefully acknowledge the following:

Chapter 4: Diane E. Bailey, Paul M. Leonardi, Jan Chong, "Minding the Gaps: Understanding Technology Interdependence and Coordination in Knowledge Work," *Organization Science* 21, no. 3 (2010): 713–730. Copyright 2010, the Institute for Operations Research and the Management Sci-

ences, 5521 Research Park Drive, Suite 200, Catonsville, MD 21228 USA. Reprinted by permission.
Chapter 5: Diane E. Bailey, Paul M. Leonardi, Stephen R. Barley, "The Lure of the Virtual," *Organization Science* 23, no. 5 (2012): 1485–1504. Copyright 2012, the Institute for Operations Research and the Management Sciences, 5521 Research Park Drive, Suite 200, Catonsville, MD 21228 USA. Reprinted by permission.
Appendix: Diane E. Bailey and Stephen R. Barley, "Teaching-Learning Ecologies: Mapping the Environment to Structure through Action. *Organization Science* 22, no. 1 (2011): 262–285. Copyright 2011, the Institute for Operations Research and the Management Sciences, 5521 Research Park Drive, Suite 200, Catonsville, MD 21228 USA. Reprinted by permission.

Introduction: Three Product Designers at Work

Arvind, in Automotive Vehicle Design

In a sparkling new technology park in Bangalore, India, Arvind, a young engineer in a large international automotive company, sits before a high-powered workstation. The technology park's five tall buildings sport marble-and-glass exteriors, emblazoned with inspirational names in large steel letters: Innovator, Discoverer, Explorer, Creator, and Inventor. Occupying the buildings are such firms as Wipro, AOL, Lucent Technologies, and Fujitsu, plus Arvind's firm, International Automobile Corporation (IAC), which is located in the six-story Creator building. The buildings on either side of Arvind's building have twelve and fourteen stories, respectively; the other two buildings sit to the rear and are also taller than Creator. Twenty-five thousand people work in these five buildings. Hidden from view is an underground mall with restaurants, banks, and gift shops.

The cubicle desk at which Arvind sits on the third floor of Creator is one of more than 120 desks in a large, open floor plan. Beige, fabric-covered walls about four feet high separate the cubicles. Each cubicle features a molded laminate desktop that extends along three of its walls; the wall juts out slightly in the middle to divide the cubicle, creating space for two engineers. This arrangement leaves every engineer with an L-shaped desktop. Each engineer sits in a task chair with wheels at the rounded intersection of the "L," facing a large computer monitor. The keyboards for the computers rest on pull-out trays under the desks. There are no bookcases and few books in any of the cubicles.

Like most of the other cubicles, Arvind's cubicle is just short of barren. To the right of his monitor sit a telephone and a water bottle; a second telephone sits to the monitor's left. On the wall behind this phone is a computer printout with an image of a vehicle; arrows lead from the various components to their names on the side. Below this printout hangs a twelve-month calendar printed on a single page. A postcard bearing IAC's mission statement and core values—"Integrity," "Teamwork," "Innovation"—is also tacked on the wall. A small stack of printouts, an open white notebook with a cell phone resting on top, a pen, a pencil, a letter from a bank (it appears to be offering a credit card), some plastic protectors for paper, and an eraser for the whiteboard complete the cubicle's inventory. The whiteboard hangs on the right wall and is full of project notes. The whiteboard eraser is barely used, its white surface merely tinged around the edges. An empty plastic inbox hangs from the wall. The dearth of work artifacts suggests that Arvind would need only a few minutes to gather his work and personal possessions if asked to vacate the office. Nothing in this cubicle is suggestive of his personal interests or hobbies, his family, or his beliefs (unless we count the IAC values).

Today, Arvind is wearing a plaid short-sleeved cotton shirt, dark pants, and tan shoes. A silver watch rests on his left wrist, a rakhi on his right. He stands about five feet, eight inches tall and has short hair and a thick mustache.

Two images fill Arvind's computer screen. On the right side of his screen is a photographic image of a physical automotive part that has undergone a controlled crash test. Through an FTP (file transfer protocol) connection, Arvind had retrieved the photograph of the crushed part from a database at the vehicle proving grounds in Australia where the crash occurred. On the left side of Arvind's computer screen is a brightly colored graphic image. This graphic image shows the same part as in the photograph, also crushed, but the crash this part endured was virtual, not physical. Arvind carried out the virtual crash on his computer in India. He did so using a simulation software application, having first built a model of the vehicle with graphic design files he received from Australian engineers. As Arvind sits in front of his computer, he looks back and forth between the two images of crushed parts, one real, one virtual. His goal is to make the image on the left look exactly like the one on the right—in short, to make virtual reality exactly mimic physical reality.

Arvind and his colleagues in Bangalore employ a combination of advanced computational software applications, high-speed information and communication infrastructures, and large databases to create, modify, and exchange around the globe a variety of digital files. Combined, these technologies permit Arvind—a well-educated and highly trained engineer—to work on state-of-the-art automotive design projects with far-flung colleagues in Germany, Sweden, Korea, Brazil, Mexico, Australia, or the United States, even though he lives in a country where most people still do not own cars. Arvind counts himself among the carless; like many of his colleagues, he has ridden in a car only a handful of times. In lieu of driving, Arvind rides a bus on a long commute to and from work that sometimes exceeds two hours.

Each evening when Arvind leaves work for home, an engineer on the second shift will come and sit in Arvind's chair at Arvind's desk in front of Arvind's computer, working until after midnight on a different set of tasks. In the morning, when Arvind returns to the office, he will grab a cup of chai in the company canteen and then reclaim his shared chair, desk, and computer to renew work on his design and analysis tasks.

Sally, in Structural Building Design

About forty miles south of San Francisco, a young female engineer named Sally works as a structural engineer at Tilt-Up Designs, a building design firm whose office is located in a business park in Palo Alto, California, on the affluent San Francisco Peninsula. Home to Stanford University and the venture capitalists of Sand Hill Road, the peninsula marks the beginning of the entrepreneurial expanse known as Silicon Valley. At the business park where Sally's firm is located, a row of tightly spaced trees separates the three nondescript, two-story cream-colored buildings from U.S. Highway 101. A directory sign in the parking lot gives the names of firms in the park, with BetaSphere, Ibrain Software, SyntelSoft, Smart Corporation, Taviz Technology, Analytica, and Thin Multimedia among them. As a firm that employs structural engineers, not computer programmers, Tilt-Up Designs stands out among these Internet and high-tech ventures as a decidedly "old economy" neighbor.

Inside the glass doors of the lobby in Building 3, a carpeted stairway leads up to the second floor. The firm's large double wooden doors, evocative of a law office, are the first on the right and open into a large room with a view of almost the entire office space. A curved piece of green glass at the reception desk and a large, curved, mango-colored wall that sets off the kitchen lend a hip, modern touch to the office.

At about ten feet square, the cubicles in the office are open and airy. The gray-colored side walls extend only a foot above the desktop, allowing engineers to easily look into a next-door neighbor's cubicle. The back wall of each cubicle is much higher, perhaps five and a half feet, so that an engineer has to stand on her toes when she wants to peer into the cubicle behind hers. All the cubicles look exactly the same: each houses an L-shaped, laminate desktop and a separate, much older, wood-and-metal drafting table. Along the back wall of each cubicle are three bookshelves in a long row, two of which are covered with metal flip-down doors with small built-in locks.

Sally's cubicle is cluttered. A set of 3′ × 4′ paper engineering drawings lies flat on top of her drawing table. A box under the table contains more drawings; these drawings are rolled up and in a range of sizes. Eight black notebooks of different widths stand in a line on the side desktop. A two-tray plastic inbox supports the notebooks on the right side; resting in the inbox are a couple of 1-inch cables with blue casings. A stack of yellow manuals, volumes 350–353 of a series titled *Steel Moment Frame Buildings,* rests atop the notebooks. With them is a notebook neatly labeled *Current Design Issues in Structural Engineering,* and a book, *Seismic Provisions for Structural Steel Buildings.* On the other side of the black notebooks are a Nerf ball, two ceramic mugs full of pens, and two photographs, one of which presumably features her husband, who she later reveals is a software engineer.

The volumes in Sally's bookshelves along the back wall of her cubicle feature textbooks from her engineering program as well as a number of design manuals and vendor catalogs. The titles include *Design of Wood Structures, Mechanics of Materials, Structural Steel Design, Reinforced Concrete Design, Soils and Foundation, Reinforced Masonry Design, UBC [Uniform Building Code], Design of Reinforced Masonry Structures,* and the *Hilti* catalog of bolts. A white construction hard hat sits at the end of the line of books.

Beneath the bookshelves, on the back desktop, are an electric pencil sharpener, a stapler, a tape dispenser, the phone, and more black notebooks. Photocopies of tables of numbers and other information line the back wall. One of them is the Tilt-Up Designs phone directory, which lists the names and phone extensions of its thirty-odd employees; it does not list their email addresses. A digital photo printout pinned to the wall shows a handful of runners, including Sally, in the Portland Marathon; the printout lists her age (twenty-four) and her race time. A second digital photo printout shows Sally on a construction site. A black shoulder bag and a pair of low hiking boots rest on the floor. A photo ID badge, issued by a prominent chipmaker (a client) and featuring Sally's photo, lies on top of the bag. Large rolls of engineering drawings, in addition to those stacked up under the drawing table, lie on the back desktop. The entire cubicle to the left of Sally's, having seemingly been converted into a storage space, houses only engineering drawings, some of them on wooden racks, like newspapers in a library.

This host of work artifacts, many of them paper-based, distinguishes Sally's cubicle from Arvind's. If she needed to move to a different cubicle, it could easily take her a full day to transport and resituate her many possessions. Yet most of the items in her workspace are work-related, with only a few personal touches—the photographs, the mugs—to distinguish it as her own. Someone else would have difficulty sharing her cubicle with her—say, working on a different shift, as in Arvind's case—because the mass of working documents on her desktop and drawing table are project-specific and bulky; they could not easily be shunted aside to make room for a different set of documents.

Sally's computer monitor is nearly hidden in this setting; it rests in the deepest corner of her cubicle in the pocket of the L-shaped desktop. Her computer tower sits below the desk. The computer is a non-brand-name system (purchased from a company called Whitebox). Some yellow Post-it notes are stuck to the desk in front of her keyboard. Sally's computer screen is dark and will remain dark until well past noon, although she arrived at work just before 9 a.m. The computer screens on the desks of Sally's thirty or so colleagues are similarly dark and tucked away, their owners huddled nearby over large paper engineering drawings, thick design manuals, and stacks of sheets bearing hand-worked calculations. For Sally and her colleagues, the workplace is overwhelmingly a physical reality, not a virtual one.

Today Sally is wearing gold wire-framed glasses, a white, short-sleeved knit top, tan dress pants, and brown dress shoes with a wide heel. She has curly shoulder-length blond hair. Her silver watch rests on her right wrist.

Sally will work at her desk today with her primary workplace technologies: her tablet of engineering grid paper for writing down calculations, her pencils, her handheld calculator, a design manual or two, and project binders full of calculations from past and current projects. These are the simple tools with which Sally most often carries out her work. She uses these tools to sketch diagrams of columns and beams on her calculation sheet, to build models, and to solve mathematical formulas so that she can design a structure that will successfully transfer building loads from one story to the next, from top to bottom and, eventually, to the ground. Sally will turn to her computer only once today, and then only for a short while. She will use her computer to conduct analyses that require complex computations via sophisticated analytical software. The analyses will tell her how her building design is likely to fare under seismic forces, such as those that accompany an earthquake. Each seismic analysis that Sally runs with the software will finish in less than a minute; a full session of testing will take less than one hour, after which she will again turn away from her computer and return to her other tools.

Each evening Sally turns off her computer promptly at 5 p.m., when she leaves the office. Her computer will remain off all night. She will not access her work from home in the evening because her firm neither supplies home computers for its engineers nor provides a network for remote access. When Sally returns to work each morning, turning on her computer is not part of her morning routine. Instead, she sips her morning coffee and tidies up her desk, organizing the books, manuals, drawings, pencils, erasers, calculator, and grid paper that she will need to accomplish her design and analysis tasks.

Eric, in Computer Hardware Design

Deep in the heart of Silicon Valley, a young hardware engineer named Eric sits in his chair in front of his computer monitor. Eric is a product designer at Configurable Solutions, a computer chip design firm; he designs microprocessor cores and their peripherals. His firm is located less than ten miles south of Tilt-Up Designs, just on the other side of U.S. 101. Housed in a low-slung, single-story building with a red tile roof, Configurable Solutions occupies one among a dozen or so identical commercial buildings in a Santa Clara business park nestled among a host of high-technology businesses just north of the San Jose airport. This area is no-holds-barred "new economy" turf, with parking lot after parking lot signaling abundant office space and commercial development.

The glass front doors of Configurable Solution's building open onto a spacious lobby with cream-colored walls and a blue-and-mauve carpet. A display case, tucked in a corner, holds plaques that recognize membership in industry associations: Artisan Partner Network, Fabless Semiconductor Association, Synopsys Interoperability Partner, and the like. A few mounted print articles, culled from the *Wall Street Journal*, *Electronik*, *Penton's Embedded Systems Development*, *Red Herring*, *Electronic Design*, and the *San Jose Mercury News*, carry stories on Configurable Solutions's technology and success.

Most engineers in the firm sit in cubicles, but Eric and his officemate got lucky: they have a glass-walled perimeter office. In it, two stand-alone desks with task chairs and a single bookcase constitute the firm-issued office furniture. Their bookcase is loaded with guides to computer languages, software, utilities, and systems (e.g., Make, Perl, GNU), vendor manuals for commercial software applications in hardware design (e.g., 0-In Check), and a couple of textbooks. The computer monitor consumes much of Eric's desktop. A phone rests beside the monitor; a split keyboard rolls out on a separate tray. The sole work-related artifact on Eric's desk other than his phone and his computer equipment is a small metal document stand holding several computer printouts at eye level beside the monitor. Behind the desk is a large whiteboard with a blue diagram sketched on it; over the course of the five months that we visit this site, the contents of the whiteboard will remain unchanged.

The remaining items in Eric's workspace are all personal items. A row of four medium-sized stuffed animals, including a ladybug, lines the top of his computer monitor. Three large wedding photographs in silver-and-oak frames decorate his desktop. Around the monitor lie a yo-yo collection, a few action figures (including Agents Mulder and Scully from *The X-Files*, a TV science fiction show), and a Rubik's Cube. A rainbow tie-dyed T-shirt adorns his chair. An enormous round white wicker chair with a cushion, large enough to take a comfortable nap in, partly blocks the door to the office. Eric's Canondale mountain bike rests against a wall.

Overall, Eric's desk, like Sally's, is a bit cluttered, but not because of his work. If he had to shift to another office, it would take him perhaps half an hour at most to place his personal belongings in a box for transfer. Although another engineer in the firm could probably work efficiently in Eric's workspace, she might not be comfortable doing so in the long term unless she shared Eric's fondness for his many personal artifacts.

Eric's outfit today—a white knit polo shirt with tan cotton shorts, white athletic socks, and Birkenstocks—speaks to his geek credentials. Asian and about five feet, seven inches tall, Eric has short black hair and wears wire-rimmed glasses.

Eric works almost exclusively in the realm of the digital, but what Eric and his colleagues examine on their computer screens are not virtual replicas of physical parts, such as Arvind sees at IAC, but the words, numbers, and symbols of line after line of programming code. Although Eric need devote no time to physical housekeeping in his office, a task that routinely occupies Sally, he will use prompt commands such as "rm" (remove) and "mv" (move) to sort, transfer, and delete digital files and directories in an effort to prune and organize the virtual clutter of program files, test results, and debugging logs that he amasses daily.

While working on his computer today, Eric will operate on multiple digital levels simultaneously. He will send his analyses, which typically take hours to finish, off to a bank of five powerful servers, where they will run "in the background." As he waits for his tests to finish, he will busy himself with other computer-based design tasks on his own computer. For example, Eric searches digital bulletin boards for small software programs that he can incorporate into his work, he reads user group threads about new technologies, he browses the website of an upcoming equipment conference, and he writes new code that he will test later.

And whereas Sally's computer is never busy when she is out of the office and Arvind's is busy even when he is away (not with his tasks, but the tasks of the other engineer who sits at his desk when he is not there), computers carry out Eric's work nearly around the clock. Eric's firm supplied all of its engineers with a home computer and a network connection so that they can work evenings and weekends. Consequently, Eric often wakes up about 2 a.m. to check on the analyses that he sent to run on the servers the evening before. Eric checks his tests because if they have failed he wants to fix them and send them to run again. That way, he will have results to work with in the morning at the office.

This 'round-the-clock cycle of work means that Eric's office computer was last turned off one and a half years ago when the power in the company's building unexpectedly went out. Neither Eric nor any of his colleagues, however, routinely works at home during the day despite having the equipment to do so, and none of their work is outsourced or offshored to distant others.

About 10 a.m., when Eric arrives at the office each morning, he walks through a maze of cubes whose five-foot high walls lend privacy and quiet to the engineers. In search of chocolate pastries, Eric often stops at the break room, where employees fuel up on espresso and granola bars. Afterward, he passes many of the eighty or so engineers in his group, each sitting, as he will soon sit, directly in front of a large computer monitor that is the dominant feature of the workspace.

Arvind, Sally, and Eric are all modern-day engineers engaged in the design of complex products, but they use advanced information and communication technologies—such as digital computers, computer networks, the Internet, email, and a host of sophisticated computational, logic, and graphical computer applications tailored to engineering design and analysis—to different extents, in different ways, and with different results. The three vignettes illustrate that these technologies play different roles in the work of each engineer; the vignettes also point to different patterns of use and different consequences. Combined, these different outcomes reflect Arvind's, Sally's, and Eric's technology choices.

- For Arvind, advanced information and communication technologies are central to his ability to carry out his everyday work; more important, without them, he could not work in India at a great distance from his colleagues around the world.

- For Sally, these technologies take a backseat to more traditional paper-based technologies in her everyday activities, and she employs none of their capabilities for working remotely or at a distance from colleagues.
- Like Arvind, Eric could not complete his everyday tasks without these advanced technologies, but his use of them to work remotely lies somewhere in between the extremes represented by Sally (not at all) and Arvind (to the limits of the globe).

Why did these product designers make such different technology choices? Why was their embrace of advanced information and communication technologies so different when their work was so much alike? Why did a single best way to use advanced information and communication technologies in this work not emerge? Is it simply a matter of time before the use practices of Arvind, Sally, and Eric converge to such an optimum? Can we attribute the differences in technology choices across Arvind, Sally, and Eric to idiosyncratic preferences across individuals?

We address these questions in this book. Their answers are important because continuous advances in information and communication technology more or less guarantee that new technologies will regularly and frequently appear in the workplace. Competition among technology producers will likely continue to yield, as it does now, a market with multiple technologies to meet the same workplace need, thus providing a range of options for managers who make technology adoption decisions. Surely cost, quality, and reliability will be among the technology features that managers will evaluate as they consider these options.

But managers may also be concerned about other technology features that are much more difficult than cost, quality, and reliability to precisely specify, quantify, and evaluate features that are reflected in the technology choices that workers make. One such feature is the role that the new technology will play in the workplace. To what extent will the technology perform a task in lieu of a worker versus simply aid in the worker's task execution? To what extent will the worker trust the technology to independently and correctly complete the task? In what ways and to what extent will the new technology support how the worker thinks through the task at hand? Other features that evade easy measure and analysis are the expected patterns of use of the technology and the consequences of that use. Will all workers use technology in the same way? Will workers' use of the new technology, be it uniform across the workplace or idiosyncratic to the worker, alter the ways in which they work? Will use practices modify the products that workers produce, the social practices in which they engage, the division of labor under which they operate, and the routines and standards they follow?

Unfortunately for managers, variation in technology choices, such as what we see among Arvind, Sally, and Eric, severely complicate the problem of predicting the impact of a new technology. If, for example, use never converges to a single (perhaps best) practice, then organizations will struggle to predict the impact of new technologies on their workers and technology pundits will have difficulty foretelling the impact of new technologies on the workforce as a whole. The immediate benefits of explaining variation in technology choices lie, therefore, in helping managers make better decisions when purchasing new workplace technologies, providing workers with a firm idea of how their work is likely to change, and guiding organizations in thinking about how to organize and equip their workforce. Benefits exist for technology producers as well, in the form of valuable feedback as they contemplate the design of their products. More broadly, explaining such variation ought to help shape the aspirations of new workforce entrants, inform educators as they prepare students for jobs and careers, and aid government policy makers in crafting national labor, education, and business development goals and incentives.

Although answering the question of why technology choices vary will consume us throughout this book, we can answer one question immediately: the differences in the extent to which, and the manner in which, Arvind, Sally, and Eric embraced advanced information and computer technologies were not simply a matter of individual preferences. We know how these three product designers employed these technologies because our research team observed Arvind, Sally, and Eric in their workplaces as the engineers went about their everyday activities. Over the course of several years, our team also observed many of Arvind's, Sally's, and Eric's organizational colleagues, and we observed product designers just like them at other sites and firms. As a result, we know that the differences we saw in technology choices across Arvind, Sally, and Eric were not unique to them. Rather, Arvind's choices were pretty typical for automotive engineers, Sally's choices were common among structural engineers, and Eric's choices were representative of other hardware engineers'. We therefore sought to understand why differences in product designers' embrace of advanced information and communication technologies seemed to fall along occupational, rather than individual or organizational, lines.

This book documents what we found. Overall, we discovered that engineers face different occupational factors, and that these occupational factors shape their technology choices. For example, the three engineering occupations represented by Arvind, Sally, and Eric create products for different kinds of markets, experience different rates of change in the bodies of professional knowledge they employ, are held to different standards of product liability, and can reduce product complexity to different extents. These occupational differences have implications for how product designers learn from and teach one another on the job, how and whether they communicate with peers across firms, how status and power play out in terms of tenure and hierarchy in their companies, how the culture of their organizations becomes formal or informal, and, underlying all of this, how, why, when, and where they embrace advanced technologies to help them in their work. In other words, these occupational factors, whose roots are largely historical, social, and economic, not technological, nonetheless shape the role of new technologies in each occupation's work.

In formulating an occupational explanation for differences in technology choices in the workplace, we challenge existing explanations. One such explanation that we challenge is the notion that one can predict choices by examining the fit between the material functions and features of a technology and the requirements of the task (including the desired outcomes) to which it might be applied. In short, arguments about fit ask, is this the right tool for the job?

The stories of Arvind, Sally, and Eric render such task-technology fit arguments suspect because these product designers, as we explain in more detail in chapter 2, were engaged in fundamentally similar design and analysis tasks. In each case, the engineers built conceptual models or physical prototypes of their design solutions to facilitate analyses and tests of performance. The outcomes of these analyses and tests routinely signified failure on one or more performance metrics, thus calling for a modification of the design and a new round of analysis. We found such cycles of design and analysis—replete with their associated tasks of problem solving, sketching, making assumptions, modeling, calculating, estimating, experimenting, and debugging—in each occupation we studied, but just because product designers in all three occupations engaged in these tasks did not mean that they employed advanced information and communication technologies in the same way or to the same extent.

Overall, the fit argument represents necessity, not sufficiency. True, the technology must work, at some level, for the task at hand. But many possible technologies might accomplish the job, and any given technology might invite a multiplicity of possible ways to do the work. Automotive engineers, for example, need not have distributed their work globally; they could have used information and communication technologies to simulate vehicles locally rather than spread their work across countries. Structural engineers, if they so desired, could have allocated more of their tasks to the computer and, in so doing, given a larger role to this technology, but the firms we studied purposefully chose not to do so. Hardware engineers could have employed information and communication technologies in the same way that Arvind and his colleagues did, distributing design work around the globe so that they could sleep through the night, but none of the engineers we studied chose that route. Insofar as the fit between the task and the technology did not limit how the product designers did their work, some other factor or factors must have shaped technology choices to yield the similarities we observed within occupations and the differences we observed across occupations. We show in this book that we can usefully group those factors according to occupation to develop an occupational perspective on technology choices.

Another explanation that we challenge is the idea that technology choice is primarily a function of social setting. We certainly do not deny that the choices individuals make about which technologies to use and how to use them may be influenced by others. University programs, for example, often shape structural engineers' choices, with professors steering students toward particular software applications, which the students then continue to employ long after graduation. In the field, we could at times guess when engineers graduated by the software application we observed them using for a given task. In general, however, no matter which specific application they employed, product designers in structural engineering used technology in similar ways and to a similar extent, with similar consequences across firms, and these ways, extents, and consequences looked quite different from those in the other two occupations. That is to say, the influence of social setting was manifest within the broader context of a stage set, we argue, by occupational factors.

In chapter 1, we review how scholars have historically explained technology choices in the workplace. Two schools of thought have dominated explanations to date. The first school, known as technological determinism, predicts inevitable and universal roles, patterns of use, and consequences for a technology; the second school, social constructivism, argues for local roles, patterns of use, and consequences that are particular to individuals or organizations. To illustrate these two schools of thought, we employ the concrete examples of the two technologies that perhaps most clearly defined workplaces in the twentieth century: the computer (which began to make its mark in the latter half of the century) and the moving assembly line (which rose to prominence in the first half of the century). Through these examples, we point out the strengths and disadvantages of technological deterministic explanations, on the one hand, and social constructivist explanations on the other.

We also consider alternatives to these two dominant schools of thought. To situate these alternative perspectives, we introduce two distinctions. The first distinction is between determinism (external forces are the agents of change; outcomes are inevitable) and voluntarism (humans are the agents of change; outcomes are not inevitable). The second distinction is between materialism (physical causes drive human action) and idealism (ideas and beliefs drive human action). We show that, in the two-by-two scheme that results from these two distinctions, technological determinism is a form of materialistic determinism, and social constructivism is a form of idealistic voluntarism. Recent alternatives to these two schools of thought, including sociomateriality and critical realism, are forms of materialistic voluntarism. We argue that idealistic determinism, the understudied fourth conceptual space in this two-by-two scheme, offers the potential for a new, alternative perspective whose advantages address the shortcomings of existing explanations of technology choice.

This new perspective, which we designate an "occupational perspective," allows for some homogeneity in technology choices across organizational contexts without subscribing to the hard tenets of materialism. By adopting an idealist stance on determinism, we can explore how external forces shape ideas, beliefs, norms, and values in ways that steer individuals to particular technology choices in somewhat predictable and generalizable ways. We acknowledge that a deterministic approach of any type diminishes the concept of human agency and thus alters a bit the notion of choice. Yet in the cases that we investigate, choices were apparent; managers and product designers who made the choices may have had little leeway in their decision, given the strong occupational factors acting on them, but in each case there were always alternative options that they might have chosen. For this reason, we retain the notion of choice despite our deterministic bent.

Our occupational perspective proves adept in explaining technology choices across the three engineering occupations that we studied. We argue that an occupational perspective is a highly viable tack for future research because occupations are recognized and accepted groupings of individuals in the work context, because special-purpose technologies are often designed with particular occupations in mind, and because occupations often speak with a single voice—as through professional organizations—when deciding matters that affect the factors that shape technology choices. At the book's end, we consider both the potential and the limitations of an occupational perspective in explaining technology choices across a broader range of occupations than the three that we studied.

In chapter 2 we describe the engineering work, technologies, and products in each of the three occupations we studied. We show that the types of tasks involved in product design in each occupation were similar, involving repeated cycles of design and analysis as engineers built, tested, analyzed, reconfigured, and tested again their models until the models achieved desired levels of performance. Additionally, similar types of technologies were available to engineers across these three occupations. In terms of general-purpose technologies, the engineers all had available to them the digital computer, computer networks, the Internet, email, word processors, text editors, spreadsheet applications, and the like. Their special-purpose technologies also displayed considerable similarity: engineers in all three occupations had available to them sophisticated computational, logic, and graphical computer applications to aid their engineering design and analysis tasks. Finally, we show that the objects of design—automobiles, buildings, and chips—were similar in terms of the degree of their complexity. Each product consisted of tens of thousands of components whose integrated functioning was essential for successful performance. Drawing on our observations of Arvind, Sally, Eric, and their colleagues, we provide numerous detailed explanations of work, technology, and product complexity to make clear the similarities across occupations. Yet despite these similarities, the engineers in our study made different technology choices. Our objective, then, was to understand why significant differences in technology choices existed across occupations so similar in their fundamental nature.

To address this question, chapters 3, 4, and 5 focus in turn on the primary relationships that are likely to change in the face of technology choices: human-technology relationships, technology-technology relationships, and human-human relationships. In each of these chapters we build narratives to explain the choices that engineers in each occupation made and the occupational factors that shaped the engineers' motivation (their ideas, beliefs, norms, and values) in those choices. In chapter 3 we focus on the role of technology in the three occupations; in chapters 4 and 5 we explore specific patterns of use. In all three chapters we consider the consequences of the engineers' technology choices.

Specifically, in chapter 3, we consider human-technology relationships by examining which tasks product designers allocated to technology and which ones they kept for themselves. The choice that consumes us in this chapter is the extent of the role that the engineers bestowed on their computers. We found that the engineers' choices varied significantly, with one occupation purposefully minimizing the computer's role, a second opting to maximize it, and the third seeking to balance it with the role of physical artifacts. We tease out why structural engineers in building design allotted the computer only tasks that demanded considerable computation or representation, why hardware engineers in computer chip design assigned to it every task they could, and why automotive engineers in vehicle design were somewhat torn over how many tasks and which kinds of tasks should be handed over to the computer.

Chapter 4 explores technology-technology relationships, looking at the extent to which product designers streamlined their design process. The specific technology choice that we examine in this chapter is whether or not engineers opted to automate the transfer of intermediate work products, a pattern of use that spoke to their willingness to extract themselves from sequences of tasks. Automation in this case involved linking technologies so that the output of one technology became the input of the next technology without human intervention. We explain why structural engineers barely streamlined their design process at all, manually entering the output of one stand-alone technology as the input for another, whereas hardware engineers took as a key objective the smooth automation of work such that linked technologies came to form a "design flow." Automotive engineers seemed headed in the direction of hardware engineering, but lagged behind; we discuss a case of their mixed motives when automating functions around a particular technology, and why their efforts fell short of their goals.

Human-human relationships are the focus of chapter 5, which asks how product designers located work in the context of a host of advanced information and communication technologies that were equipped to digitize and transport work artifacts across distance and allowed shared access to remote depositories. Specifically, we examined whether or not engineers employed these technologies so that they might work at a distance from one another via telecommuting or offshoring, patterns of use that spoke to the engineers' willingness to abandon proximity to one another. It may come as no surprise that the occupation that allotted the smallest role to the computer and automated the fewest links between technologies also kept all of its workers under one roof (structural engineers); in this field, side-by-side human action and interaction reigned supreme. But why did hardware engineers make the same choice even though so much of their work was accomplished using a computer? Automotive engineers, who were the least sure about which tasks they should allocate to the computer and the extent to which they should automate links between technologies, chose to distribute work (and workers) around the globe. We explore these choices in this chapter

Our thesis in this book is that differences in technology choices that arise in human-technology, technology-technology, and human-human relationships derive in large part from occupational factors. In chapter 6 we bring together the threads of those factors from the narratives told in earlier chapters. In total, we consider twelve occupational factors that emerged from our data. Three of those factors concern the environment: market, competition, and professional liability/government safety regulations. Three factors arose in the context of knowledge: the rate of knowledge change, testing capability, and self-explanation of work artifacts. Two factors center on work organization: the division of labor and task interdependence. The product was the focus of two other factors, product interdependence and complexity reduction ability. Finally, technology itself contributed two other factors, cost and transparency.

In chapters 3, 4, and 5, we show how different combinations of these twelve occupational factors strongly shaped the technology choices that engineers in our study made, explaining why, how, and to what extent engineers embraced advanced information and communication technologies. In chapter 6 we consolidate our findings from the earlier chapters to build single narratives for each occupation. In other words, we look across relationships within each occupation in an attempt to build a holistic understanding of which factors shaped engineers' technology choices in that field. We conclude chapter 6 by shifting our focus away from the three occupations to the factors themselves, to tentatively explore each factor's independent predictive capability. In this manner, we investigate whether it might be possible to narrow the focus to just a few key occupational factors rather than considering, as we did, the particular combination of factors within an occupation that appeared to jointly shape each technology choice.

Chapter 7 concludes with the benefits, implications, assumptions, and limitations of the occupational perspective we develop in this book. An occupational perspective promotes an understanding of past technology choices and provides leverage in predicting future ones. For example, at the same time that an occupational perspective allows us to understand why

Arvind, Sally, and Eric made different technology choices, it also points to the fields in which we might expect to find more people like Sally and Eric, whose jobs seem secure in their U.S. locations, and the fields in which we can increasingly expect to find people like Arvind, who can carry out his work halfway around the globe from his working colleagues. An occupational perspective also helps us to recognize how certain occupational phenomena, such as work culture and the physical environment, derive from technology choices. Because the occupational perspective that we propose rests on the twelve factors that emerged from our data, we consider the extent to which our findings are dependent on our sample of three engineering occupations—that is, we consider the generalizability of our work. We also address the possibility that advances in technology might render our findings no longer accurate, and we discuss other possible limitations and assumptions of our work.

The appendix lays out our research design and our methods of data collection and analysis. We build the arguments in our book on rich qualitative data, which our research team of twenty-seven individuals collected through ethnographic techniques of field observations and interviews of engineers. Our sites included three structural engineering firms and three hardware engineering firms in the San Francisco Bay area, as well as three automotive engineering groups spread across eight global sites of a single large manufacturer. Table I.1 lists the firms' names and market focus.[1]

In total, we logged more than a thousand hours of observation of more than one hundred engineers at work and wrote thousands of pages of field

Table I.1
Firm Names and Market Focus

Occupation	Firm/Group Name	Market Focus
Automotive engineering	International Automobile Corporation (IAC) • Body Structures • Noise & Vibration • Safety & Crashworthiness	Cars and trucks
Structural engineering	Seismic Specialists	Civic building retrofitting
	Tall Steel	Multistory commercial projects
	Tilt-Up Designs	Semiconductor fabrication plants
Hardware engineering	AppCore	Microprocessor architectures and cores
	Configurable Solutions	Customizable microprocessors
	Programmable Devices	Programmable logic devices

notes documenting what we observed. In addition, we brought hundreds of work artifacts back with us from the field for further study. The appendix describes how we organized and managed our research.

Perhaps most important, the appendix describes the novel techniques of data collection and analysis that we developed to capture the nuances of highly technical work. Simply being a skilled observer was insufficient for documenting what engineers did all day. Because the engineers we followed used a technical language specific to their specialty, carried out tasks with few everyday correlates, used sophisticated technologies whose function was not easily gleaned, and often employed multiple technologies simultaneously, we needed new methods to understand why engineers made the technology choices they did at work and why they acted as they did. To begin, we needed to gain scientific and technical knowledge that many scholars of work lack. Next, we needed creative ways to record in the field what engineers said and did, as well as ways to later weave together recorded streams of technology use, artifact creation, dialogue, and other action. Finally, we needed ways to help us make sense of and analyze the incredibly rich data we collected. In addition to detailing the novel methods for data collection and analysis that we developed, our appendix provides suggestions for how future researchers might carry out similar investigations.

1 Explaining Technology Choices in the Workplace: Proposing an Occupational Perspective

Few technologies provide a better starting point than the computer when one wishes to explain technology choices. The roles that computers play in formal organizations, as well as the patterns and consequences of their use, have been a subject of debate since their entry into the workplace. From the early 1950s, when large companies like IBM and AT&T first implemented mainframe computers, through the 1980s, when the personal computer arrived on desktops across the country, to recent years, when laptop computers and other mobile devices such as tablets and smartphones became ubiquitous among workers on the go, academics, journalists, and workers themselves have given varying accounts of the computer's role and use. These accounts have often featured the labor, organizational, and social consequences of computerization.

Such consequences are the understandable result of new technology implementation: as a new technology settles into its role, it is likely to nudge, alter, displace, or usurp other technologies—and sometimes workers themselves—whose functions share the domain the new technology comes to occupy. In turn, existing technologies and human workers shape and often restrict the domain allotted to the new technology. Two schools of thought have dominated the discussion about the relationship between technologies and humans that emerges from this process.

The first school of thought goes by the name "technological determinism." Technological determinists argue that the role that a new technology comes to fill, its pattern of use, and its consequences are common to all settings. According to this view, the features of the technology itself drive or determine this commonality. For example, from the perspective of technological determinism, the cells and mathematical functions of spreadsheet applications afford calculation, thus driving a common use of this technology in the display, manipulation, and tracking of numbers, as in formulating budgets and tracking expense records. No one uses the spreadsheet application to write letters, surf the Internet, or crop images because its affordances lie in calculation, not in those other domains. By predicting a single, universal role, a single pattern of use, and a single set of consequences, technologically deterministic accounts of new technology introduction downplay, if not completely disregard, the impact of contextual factors such as the state of the local labor market, the history of technology adoption in an organization, or an organization's existing technology suite as alternative explanations for the new technology's effects. For technological determinists, these purely local factors lack the power to alter in any significant way the outcomes associated with a new technology.

The second school of thought, which arose in opposition to the determinists' creed, advances "social constructivism" as an explanation for the adoption and use of a particular technology. Social constructivists contend that local, contextual factors and the actions of local actors so strongly shape the process of new technology introduction that the final role a technology takes on, its patterns of use, and its consequences are tailored to each setting. The confluence of these local factors and actions shapes the outcomes associated with the use of a new technology; the outcomes are not determined in advance by any particular feature of the technology or by a shared goal of its implementers or users. A social constructivist account would explain, for example, why one firm might implement an inventory and ordering system so as to share stock levels, planned production, and manufacturing performance data with suppliers up and down its supply chain in an effort to unite its suppliers in the common goals of high customer service and low inventory carrying costs, while another firm might implement the same system but, lacking sufficient trust in its suppliers, deny their access to sensitive data concerning product quality, due dates, and the timing of special customer orders. For social constructivists there is no single, universal role, no single pattern of use, and no single set of consequences driving the adoption of a new technology in the workplace; rather, multiple roles, use patterns, and consequences may arise in response to variety in organizational settings and situations.

For technological determinists, predicting the role that a new technology will play, its patterns of use, and its consequences is largely a matter of observing the technology's introduction at a single site; all other sites will reflect similar outcomes. For social constructivists, prediction is a much more difficult affair because social constructivism holds that a unique confluence of multiple varying factors shapes these outcomes for a new technology at a given site.

In this chapter, we discuss these two schools of thought and their ability to explain and predict technology choices in the specific context of two technologies whose introduction altered the face of work for countless workers in the twentieth century: the computer, which came into widespread use in the latter half of the century, and the moving assembly line, which was broadly implemented in the first half of the century. By including the moving assembly line in our analysis, we make clear that the arguments of technological determinism versus social constructivism extend beyond, and did not originate with, the computer, which nonetheless remains central to our interest at the beginning of the twenty-first century. The moving assembly line also affords a historical parallel useful for analyzing the adoption and use of the computer. Ultimately, we point out the limitations of both schools of thought, technological determinism and social constructivism, and note that they are not the only explanations for technology choices.

To guide our understanding of alternative explanations of technology choices, we suggest that scholars have often mistaken social constructivism as technological determinism's opposite. The difference between the two schools is more complicated than that characterization would imply, however. Drawing on the work of the organization scholars Paul Leonardi and Stephen Barley, we show that two distinctions are simultaneously at play in explanations of technology choices.[1] The first distinction is between determinism and its true opposite, voluntarism. Determinism holds that outcomes are inevitable because external forces, such as technology, shape them. Voluntarism holds that outcomes are not inevitable but arise as the result of the (voluntary) action of humans. The second distinction of interest is between materialism and idealism. Materialism contends that physical causes drive human action; idealism credits ideas and beliefs with driving human action.

With these two distinctions in mind, we posit that technological determinism is a form of materialistic determinism: from the perspective of technological determinism, outcomes are inevitable; technology causes them and, in so doing, shapes human action. Social constructivism, by contrast, is a form of idealistic voluntarism: from the perspective of idealistic voluntarism, outcomes are not inevitable; humans shape them, acting in accordance with their ideas and beliefs. The two-by-two matrix of determinism/voluntarism and materialism/idealism yields two other possibilities: idealistic determinism and materialistic voluntarism. Although technological determinism and social constructivism have dominated discussions of technology choices, scholars also have developed perspectives that fall into these other two camps. We provide examples in this chapter and discuss the benefits and disadvantages of each perspective.

Because all existing perspectives have shortcomings that hamper considerably the explanation and prediction of technology choices, we introduce in this book a new, occupational perspective according to which occupational factors shape individuals' technology choices. Examples of occupational factors include product liability concerns, the rate at which knowledge in a field changes, the division of labor within an occupation, and the form of competition facing the occupation. These factors refer not to the physical environment of the occupation so much as to the ideas and beliefs that undergird the occupation. Thus, an occupational perspective takes an idealist stance on determinism.

By taking an idealist stance on determinism, an occupational perspective avoids both the constrictions of technological determinism, which dictates a single, universal role, pattern of use, and set of consequences for a new technology (allowing no variation across contexts), and the laxity of social constructivism, which specifies a unique role, pattern of use, and set of consequences for each setting (predicting variation across contexts and crediting chance for any similarities that arise). An occupational perspective holds that, whereas both technology features and the actions of practitioners contribute to outcomes for each technology, occupational factors shape the larger setting in which these forces act.

In short, this new perspective categorizes contexts along occupational lines, which, by virtue of their immediate meaning in everyday work practices, are a natural place to look when investigating technology choices. As we discuss, an occupational perspective is not the only alternative to technological determinism and social constructivism, but it may be one of the more viable ones. An occupational perspective helps explain and predict technology choices because occupations, in addition to being natural groupings at work, are often the focus of special-purpose technologies. Moreover, occupations often speak with a single voice when deciding matters that affect the factors that shape technology choices.

The Technological Determinists' Explanation

In 1958, the organizations scholars Harold Leavitt and Thomas Whisler published an influential article in the *Harvard Business Review*, which they futuristically titled "Management in the 1980s."[2] The article provides an excellent introduction to the technological determinists' school of thought.

Just four years earlier, General Electric Corporation had implemented the country's first commercial computer-based information system for processing payroll. Similar computer-based systems were beginning to appear in large U.S. corporations, leaving observers to wonder what role the systems would take in, and how they would affect, the business enterprise.

Among the many technologically deterministic predictions that Leavitt and Whisler made in their article, the most poignant was their expectation that the introduction of computer-based information systems would lead to the centralization of organizational decision authority. The new computer-based systems, Leavitt and Whisler contended, would serve as easily accessible and shared spaces for the aggregation and storage of information. By making use of this new resource, senior-level managers could know better what others in the organization knew. As a result, senior managers could make many decisions themselves that previously they were forced to leave in the hands of lower-level managers in various parts of the company. The implications were obvious: less input from subordinates into top management decisions and, overall, a more oligarchical organizational form.

Leavitt and Whisler's contention was representative of a "contingency theory" of technology choice. As the organizations scholar Joan Woodward articulated after having discovered that differences in type of production systems explained considerable variance in her data on the structure of British manufacturing firms, at the core of contingency theory was the idea that "different technologies imposed different kinds of demands on individuals and organizations, and that these demands had to be met through an appropriate organization form."[3] In other words, technology shapes organizational structure. As the organizational sociologist Charles Perrow famously put it, "technology is an independent variable, and structure … a dependent variable."[4] Moreover, contingency theorists considered each technology to have its own best organizational structure or form. In the case of computer-based information systems, that form, according to Leavitt and Whisler, was oligarchical, based on a centralization of decision-making authority.

Over the years, many scholars undertook to test Leavitt and Whisler's centralization hypothesis, and many who tested it found supporting evidence.[5] As one example, Hak Chong Lee, in a study of the computerization of work in three organizations, found that centralization occurred after implementation, though Lee was unable to determine whether the changes resulted from the introduction of the computer or from the feasibility study done prior to implementation.[6] Other scholars, however, found contradictory evidence. Jeffrey Pfeffer and Huseyin Leblebici's study of thirty-eight manufacturing firms, designed to determine whether there was a causal relationship between computerization and decision authority, is an example. After controlling for both size and environment, the authors concluded that use of more elaborate computing technology was positively associated with more decentralization.[7]

Such contradictory evidence led not to a revolt against technological determinism in general or contingency theory in particular but to an equally universal counterhypothesis: computer-based information systems would lead to *de*centralization of decision authority.[8] According to this hypothesis, the new computer-based systems would serve as information hubs not just for senior managers but for lower-level managers as well. By accessing the systems, lower-level managers could come to know what their counterparts in other units knew and thus would not need to elevate decisions to the senior managers to whom the various lower-level managers reported. Instead, lower-level managers could make decisions themselves. Today, more than fifty years after Leavitt and Whisler's important article, researchers still debate whether computer-based information systems bring centralization or decentralization in managerial decision making.

What is interesting about this debate about centralization versus decentralization is not which side has it right and which side has it wrong but why this topic remains in vogue. Debates around the outcomes of computerization arguably have staying power because people hunger for deterministic predictions. As the historian of technology Langdon Winner has argued, people desire general claims about the role that technology will fill and the kinds of changes it will bring about because such claims provide a sense of clarity about an otherwise uncertain future.[9] If senior managers know, for example, that computers will serve as information hubs, and that their own use of these hubs will centralize decision making, they can devise strategies for how to structure their workforces and can justify these strategies to boards of directors and shareholders based on the logic of universal, predictable outcomes.

Whereas the focus of contingency theorists was organizational structure and form, the focus of other scholars has fallen closer to the level of the work itself. In the quest for simple rules that would predict universal outcomes of computer technologies, for example, the labor economists Frank Levy and Richard Murnane argued that one could predict and explain the impact of new computer technologies on work through a simple two-step analysis. First, one needed to evaluate the nature of the work to be done; second, one needed to compare the strengths of humans versus computers in accomplishing that work.[10] Specifically, if one could explain the work in terms of rule-based logic, then managers were likely to adopt the computer as a substitute for human labor because computers excelled at such logic in comparison to humans. Otherwise the computer would merely complement humans in the performance of the work.

Levy and Murnane offered the examples of open-outcry bond futures pit traders and cardiologists to illustrate their formulation. Because one could fully describe the main task of the pit traders (matching buy and sell orders) in step-by-step rules, computers could take over the work and thus replace many traders. By contrast, because cardiologists' work involves pattern recognition and complex thought processes (as, for example, when they interpret images of the heart on echocardiograms), computers could not similarly take over cardiologists' work, but computers could complement it (after all, computerized imaging produced the echocardiograms in the first place).

The logic put forward by Levy and Murnane was not new. Human factors engineers and complex systems designers had been considering how to allocate functions between man and machine based on a comparison of their relative talents ever since Paul Fitts, in a landmark 1951 report, created a two-column list, one column labeled "Man" and the other labeled "Machine," and recorded the system functions at which each was superior.[11] Although Fitts lists, as they came to be called, evinced an appealing clarity in their seemingly objective evaluation of traits and their fact-based assignment heuristic, they were quickly recognized as problematic. Among the problems associated with Fitts lists was their complete disregard for the social, economic, and political factors that often dictated technology choices about when human labor versus machines would be employed.[12]

The deterministic view that a given technology determines its outcomes and use independent of social, economic, and political factors is not unique to the story of computerization. Rather, the history of new technology implementation has long been dogged by such notions. As an example, we may consider the case of the moving assembly line.

If the computer was the defining workplace technology of the second half of the twentieth century, the assembly line was its counterpart in the first half. When Henry Ford first introduced the assembly line in the flywheel magneto department of his Highland Park, Michigan, plant on April 1, 1913, he altered the face of automobile assembly work forever. Before that day, each man in the flywheel magneto area joined sixteen magnets, sixteen bolts, and various supports, clamps, and miscellaneous parts into a complete flywheel magneto assembly before starting his tasks anew on the next subassembly.[13] After the introduction of the assembly line, the workers' role was drastically smaller. The men in the flywheel magneto area now performed only one simple task each, such as placing a part in the assembly or tightening nuts to hold it in place, before pushing the subassembly down the line to the next worker. Soon chains and conveyor systems controlled the movement of subassemblies along the line, thereby dictating the pace at which men worked.

Two critical films in the 1930s, Charlie Chaplin's *Modern Times* and the French movie *A nous la liberté*, painted the moving assembly line as the source of workers' discontent, emphasizing the fractionated tasks and endless tedium associated with it. Formal confirmation of the films' claims, however, did not come for another twenty years: in 1952, the labor relations scholars Charles Walker and Robert Guest carried out what became a classic study of technology and work.[14] Although Walker and Guest posed no direct questions about the moving assembly line, nearly every factory worker they interviewed spoke about its effects on his work role, his relationships with co-workers, and his cognitive and emotional states. In interview after interview, the men reported the line to be a highly undesirable aspect of their jobs. Specifically, workers did not care for the forced pace or the mindless repetition that the moving assembly line engendered.

In the same year that Walker and Guest published their book, the now famous candy factory episode on the *I Love Lucy* television show implanted the image of forced pace on a moving line firmly if amusingly in the nation's consciousness. Stationed along a candy factory conveyor belt and desperate not to fall behind its increasing pace for fear of being fired, Lucy and her friend Ethel crammed candies into their mouths, hats, and uniform blouses. At last, when the mounting stream of candies exceeded what the two women could possibly conceal, Lucy conceded, "Listen, Ethel ... I think we're fighting a losing game!"

For the next thirty years, few people challenged the idea that the introduction of an assembly line spelled disheartening conditions and a losing game for workers. Taking the argument to its extreme, the sociologist Ely Chinoy deemed the assembly line complicit in the death of the American Dream for automotive workers.[15] Chinoy argued that men on automotive assembly lines, faced with fragmented and deskilled jobs, little opportunity for advancement, and no hope of amassing enough capital to launch a business of their own, soon came to realize that their prospects for success were few. A generation later, Ben Hamper's 1991 *Rivethead,* a firsthand account of life as a riveter in a Flint, Michigan, automotive plant, affirmed Chinoy's claims and stood as poignant testimony to how working on a moving assembly line continued to dash the dreams of the young.

The histories of the computer and the moving assembly line are awash in technologically deterministic accounts of the outcomes of technology choices. By postulating inevitable outcomes, these accounts enable scholars and pundits to make clear, straightforward predictions about the future of work and workers. Armed with such predictions, managers could confidently hire and train their workforce, organize productive activities, and fashion workplace policies. Yet despite their strong appeal, these accounts run into problems when confronted with variation in technology's role, patterns of use, and consequences.

No technologically deterministic account can explain, for example, the different roles that computers played in the work of the engineers we met in the introduction, Arvind, Sally, and Eric. These three product engineers worked with advanced information and communication technologies to accomplish tasks of design and analysis, but the role these technologies played varied considerably across the different engineering occupations, as did patterns of use and consequences. The cases of Arvind, Sally, and Eric thus undermine the position that the outcomes of technology choices are the same across diverse settings. Arvind, Sally, and Eric are not the first workers whose situations cast doubt on technological determinism; the recognition of others before them led many researchers to abandon this account in favor of explanations that underscored the importance of local contexts and actors.

The Social Constructivists' Rebuttal

The 1990s saw a turn toward social constructivism in the case of the moving assembly line with the publication of *The Machine That Changed the World,* an influential book that spread news throughout the United States and the rest of the world of Japanese automotive production methods that differed radically from traditional setups.[16] The result of an MIT study of the global automotive industry, *The Machine That Changed the World* refuted technologically deterministic accounts by challenging the equation of moving assembly lines with dismal work conditions. The book described how teams of Japanese workers in Toyota factories underwent analytical training in topics such as statistics and quality control, and then helped solve production and equipment problems on the line. These problem-solving responsibilities greatly enriched assembly line jobs, resulting in highly motivated Japanese workers and contributing to high-quality Japanese vehicles.

Japanese-run plants in North America quickly became showcases for the Toyota method of worker participation and organization, causing pundits to consider that moving assembly lines were perhaps only as bad as management allowed them to be. A case in point was New United Motor Manufacturing, Inc. (NUMMI), the GM-Toyota joint venture in Fremont, California. Paul Adler, who studied NUMMI in depth, concluded in a *Harvard Business Review* article that the NUMMI plant proved that a moving assembly line need not portend death to creativity and engaging work.[17]

Adler noted that workers at NUMMI, like their counterparts at traditional auto plants, performed fractionated tasks along the line. What made NUMMI workers different from other American autoworkers was that they analyzed, designed, and ultimately standardized their jobs themselves. The NUMMI workers also suggested improvements to the production process, rotated jobs, made decisions as part of teams, and studied statistical quality control techniques. Through this expanded role for workers, NUMMI restored to the assembly line a degree of worker autonomy and intrinsic motivation that managers had stripped away under Henry Ford's formulation. In doing so, NUMMI epitomized the Toyota method's refutation of technologically deterministic accounts of moving assembly lines: moving assembly lines could play a leading role in production without generating insufferable conditions for workers. Rather, what decided whether assembly line work was demeaning or rewarding was how managers implemented the technology.

Further subverting the axiom that the outcomes of technology choices were universal was evidence from other nontraditional automotive plants, among them GM's Saturn plant in Spring Hill, Tennessee.[18] In the Spring Hill plant, workers engaged in a broad array of offline activities, with levels of autonomy that were greater than in Japanese plants. For example, Saturn workers participated in new car and equipment design, which Japanese workers did not. The most striking change that Saturn's management implemented, however, was alteration of the configuration of the assembly line itself. Whereas assembly lines in Japanese plants were largely linear affairs that resembled those in traditional plants, with the sole addition of mechanisms that allowed workers to shut down the line, lines at Saturn consisted of wooden "skillets," or platforms, that sat perpendicularly on conveyor belts. The skillets were big enough to hold the car and the worker. The addition of skillets meant that workers could stand still while they completed their tasks, traveling with a car down the line, rather than having to keep pace beside a vehicle as the line inched it forward. The new configuration facilitated longer task cycles, allowing workers to complete a larger and hence more meaningful set of tasks than on a conventional line.[19]

Technological determinism in the context of the computer similarly came under fire when variation emerged in the role that the new technology played. As in the case of the assembly line, scholars often adopted, implicitly or explicitly, a social constructivist perspective that credited managers' choices with this variation. For example, the labor scholars Eileen Appelbaum and Peter Albin in the late 1980s noted that managers in the insurance industry made decisions about the role of emerging computer technology—in particular, online processing capabilities—to elicit a range of outcomes for workers and organizations. In some cases, managers reduced decision making to rules executed by a computer, leaving behind primarily routinized data entry jobs for humans. In other cases, managers assigned routine work to the computer and created new, highly skilled positions for workers, such as customer service representative and claims representative.[20]

The Harvard professor Shoshana Zuboff has similarly argued that managers' individual choices regarding patterns of use shaped the impact of the introduction of advanced computer and information technology in paper pulp mills.[21] With the new technology in place, workers had to replace the immediate knowledge they once had gained from walking the floor and working directly with the mill equipment—for example, judging water density by placing a hand on a paper roll or kicking a malfunctioning machine to get it working properly—with an explicit, scientific understanding of the production process to properly monitor equipment when remotely stationed in a control room. Workers thus shed their tactile and tacit understanding in favor of an abstract thought process in which they weighed production information displayed on computer monitors. These changes occurred at each plant that Zuboff studied, with workers' skills shifting from action-oriented to intellective. Managers, however, some of whom fretted over their potential loss of control, had the power to restrict the range of decisions that workers could make with the new system, thus limiting the extent of workers' skill increase. In short, although the system itself facilitated a shift toward greater intellective skills, managers' implementation choices accounted for variation in the new technology's impact on workers.

Other scholars put forth the idea that it might not be just manager's choices that prompted variation in a technology's outcomes but also agency on the part of the individuals who most directly employed the technology. In a study of the introduction of computed tomography (CT) scanners in the radiology departments of two hospitals, the organizations scholar Stephen Barley found that, despite similarity in the two settings, the technology played different roles in shaping new skills and social standing among workers.[22] In both settings, the new CT technology altered established patterns of action and roles between radiology technicians (whose job traditionally was to run the equipment that produced radiological images) and the radiologists (who interpreted the images and made diagnoses). But the new patterns reflected unanticipated (i.e., not manager-designed) differences across the two hospitals in terms of the extent to which decision making remained centralized in the role of the radiologist versus decentralized, with the greater inclusion of technicians. Drawing on Anthony Giddens's structuration theory, Barley contended that social structures were both the medium and the outcome of local practice: the case of the CT scanner implementation, Barley argued, demonstrated that the introduction of new technology, by disrupting patterns of action, could provide (but did not dictate) an occasion for changing the social structure of an organization.

The organizations and information systems scholar Wanda Orlikowski subsequently expanded the idea of structuration processes at play in technology implementation, arguing that technology is engaged in a dualistic relationship with humans. People shape a given technology during its development, implementation, and use, but with the passage of time, technology may facilitate or constrain their actions after they have reified its features. In a study of the implementation of computer-aided software engineering technology in a large firm, Orlikowski highlighted the criticality of the actions of the firm's information technology consultants who employed the technology and the consultants' agency in shaping the role that the new technology came to play.[23] With time, though, the consultants' use of the technology became institutionalized, such that new consultants, who had never experienced design without the technology, readily accepted its constraints and understood their work only in the context of the technology.

Orlikowski later shifted this focus on structuration—the reciprocal effects of technology use on organizational structure and of organizational structure on technology use—to a focus on practice.[24] In her elaboration of a practice theory of technology use, she gave the work of technology developers and designers no particular privilege or force in explanations of technology choice. Practice theory is largely unconcerned with whether people use, reject, or misuse a technology's specific features as the developer or designer may have intended. In Orlikowski's terms, practice theory "starts with human action and examines how it enacts emergent structures through recurrent interaction with the technology at hand."[25]

Accordingly, the majority of studies that applied practice theory followed a two-step process. First, these studies sought to uncover the ways in which users called forth specific features of a technology through repeated use of the technology in recurrent social practice. Second, these studies explored how those specific features changed the existing work practices of individuals within an organization. For example, the information systems scholars Marie-Claude Boudreau and Daniel Robey, in a study of a government agency's decision to use an enterprise resource planning (ERP) technology, found that the technology prompted unanticipated episodes of learning, improvisation, and the emergent reconstitution of organizational structure.[26] Boudreau and Robey showed that each time users employed the ERP technology in a new way, their work practices changed slightly, and most often unpredictably. Over time, gradual changes in work practices produced alterations in the social structure of the government agency. In this manner, human agents made numerous small improvisations in how they used the technology, which resulted in changes to the agency's organizational structure, which in turn became reified in new work practices.

As these examples suggest, a major problem with a social constructivist perspective in any of its forms is that researchers who adopt this perspective cannot say much about why a given new technology plays one role, prompts one pattern of use, and engenders one set of consequences in one organization but has a different role, exhibits different patterns of use, and has different consequences in other organizations, beyond simply saying that the differences derive from the unique character of each organization.

A social constructivist stance similarly has difficulty explaining why a given technology choice may yield similar outcomes across organizations that differ considerably in major attributes such as clients, markets, and firm size.[27] For example, the structural engineers in two firms that we studied employed advanced information and communication technologies in a manner strongly similar to the way Sally and her colleagues did, routinely opting for calculators and reference books over computers. Yet Sally's firm focused on the design of one-story fabrication plants for the semiconductor industry, whereas the other two firms focused on seismic retrofitting of government buildings and multistory commercial projects, respectively. Similarly, hardware designers in two computer chip design firms that we studied checked builds in the middle of the night, just as Eric and his colleagues did, even though one firm created microprocessor architectures and cores, the other designed programmable logic devices, and Eric's firm developed customizable microprocessor cores and peripherals. The three firms also differed in their size, ranging from 200 to 2,000 employees. A social constructivist explanation would have no alternative but to chalk up these similarities across firms to chance, a conclusion of the sort that began to trouble scholars, causing them to seek alternative perspectives.

Alternative Perspectives for Explaining and Predicting Technology Choices

Technological determinism and social constructivism are not opposites; thus, one cannot posit that alternatives to them lie somewhere in between them. Instead, as table 1.1 indicates, we can place alternatives to these two dominant perspectives in the two-by-two matrix formed by the dimensions of determinism/voluntarism and materialism/idealism. As Leonardi and Barley noted, "the distinction between determinism and voluntarism is orthogonal to the distinction between materialism and idealism; yet, social scientists frequently write as if materialism implies determinism and idealism implies voluntarism."[28] By untangling these distinctions, we can situate less-studied alternatives along the diagonal of the shaded cells in table 1.1.[29]

In the upper right quadrant of this two-by-two matrix, the first alternative, sociomateriality, represents an attempt to bring notions of materialism into social constructivist studies of organizing. Sociomateriality's origins lie in the sociology of science, where scholars for the past thirty years have argued that attending to agency and social dynamics is not incompatible with an appreciation for the material constraints and affordances that technologies bring with them.[30] The "not incompatible" position came about in response to the sociology of science's turn toward social constructivism in the 1980s, in concert with a similar turn in research on technology.[31] The rise of social constructivism had led some scholars to favor explanations

Table 1.1
Mapping Theoretical Perspectives on Technology Choices along Two Dimensions

	Determinism *Outcomes are inevitable; external forces are the agents of change*	**Voluntarism** *Outcomes are not inevitable; humans are the agents of change*
Materialism *Physical causes drive human action*	Technological determinism • Contingency theory • Fitts lists	Sociomateriality Critical realism
Idealism *Ideas and beliefs drive human action*	Deskilling/Upskilling theories	Social constructivism • Structuration theory • Practice theory

of change that privileged social over material practices. In turning the tide back toward the material and away from the social, the sociology of science scholar Andrew Pickering argued persuasively that physical phenomena resisted scientists' efforts to manipulate them.[32] This resistance was part of the ongoing conversation between scientists and objects that routinely led scientists to alter their methods and their theories. Scholars such as Joan Fujimara and Ian Hutchby further cautioned that a social constructivist orientation might be misguided because material phenomena, whether natural or technical, did things that one could not attribute to social practice.[33]

Drawing on this research in the sociology of science, Orlikowski in 2007 articulated a similar critique of social constructivist studies of technology choice. She recommended that researchers adopt a "sociomaterial" perspective to replace social constructivism. According to Orlikowski, a sociomaterial perspective "asserts that materiality is integral to organizing, positing that the social and the material are *constitutively entangled* in everyday life. A position of constitutive entanglement does not privilege either humans or technology.... Instead, the social and the material are inextricably related" (emphasis in the original).[34] In other words, sociomateriality represents an improvement over social constructivism because the latter ignores the fact that technologies can do certain things. Moreover, sociomateriality, unlike strong variants of social constructivism, recognizes that technologies are integral parts of social action, not simply things people use when interacting with others. Currently, scholars are engaged in important conceptual work to tease out sociomateriality's tenets, as well as its promise for explaining technology choice.[35] Because of the philosophical complexity of the idea, these efforts are taking time. Consequently, few empirical demonstrations of a sociomaterial perspective exist.

Critical realism offers a second perspective in materialist voluntarism. Information systems scholars, among them Alistair Mutch, Olga Volkoff, Diane Strong, and Michael Elmes, have observed that a critical realist perspective is much more amenable to empirical research on technology choice than is sociomateriality.[36] Critical realism is a philosophical stance that recognizes the potential existence of a reality beyond our knowledge or conscious experience.[37] The philosopher D. C. Phillips summarized this stance as "the view that entities exist independently of being perceived, or independently of our theories about them."[38] Most critical realists hold that although humans cannot directly observe mental states and attributes (such as meanings and intentions), these mental states and attributes are part of the real world. In the context of technology choices, critical realism (in contrast to sociomateriality) holds that the social and the material are separate entities: the social and the material simply appear inseparable because human activity joins them over time. Thus, the crux of the difference between sociomateriality and critical realism is that the former treats the sociomaterial as something that exists prior to our perceptions of it while the latter argues that the social and the material are independent entities that become "sociomaterial" when human action relates them.[39]

In sum, sociomateriality and critical realism, the two perspectives in the upper right cell of table 1.1, which represents materialistic voluntarism, are recent alternatives to technological determinism and social constructivism as explanations of technology choice. In the literature, discussions of both perspectives remain primarily at the conceptual level; how either perspective might fare empirically is an open question.[40] The philosophical complexity behind these perspectives suggests, however, that they may be difficult to translate into pragmatic operational strategies to explore empirically their potential to explain technology choices. At this time these two alternatives are intriguing and promising, but their potential is largely theoretical and unproven.

The other existing alternatives to technological determinism and social constructivism are deskilling theory and its companion, upskilling theory, which appear in the bottom left cell of table 1.1. These theories represent idealistic determinism. In their fixation on technological determinism and social constructivism, scholars of technology choice have tended to overlook this long-standing pair of theories, which arose in the context of the growing use of computers at work. For workers and observers of technological change who viewed the computer as the latest installment in a long line of factory and office automation technologies that, step by step, would replace workers with machines, these theories gripped their attention because the specter of worker deskilling was among the most chilling of deterministic predictions surrounding this new technology.[41] Steeped in decades of investigative reports and academic field studies, these theories carry the weight of empirical evidence that is missing for perspectives in materialistic voluntarism.

Fears of deskilling found early support in evidence from a variety of service establishments in which the implementation of computers had removed much of the need for workers' intuition and intellect. Through interviews with workers, the writer Barbara Garson discovered that computers at McDonald's restaurants told fry cooks when to yank the fries out of the vat and told managers how many workers to place on the schedule.[42] Cashiers at the burger chain no longer needed to remember prices, to add sums, or even to ring up numbers: keys on the computerized cash registers showed pictures, not prices, of menu items, rendering the register's user interface not unlike a Fisher-Price See 'n' Say toy. Reservation agents at American Airlines described to Garson how computerized reservation systems monitored how long they spent on each call, how many calls they converted to bookings, and how much time they spent at their desks "unplugged" from the phone. The information collected by the computers enabled managers to standardize agents' phone conversations with customers. The typical two-minute reservation booking call was divided into four parts—opening, sales pitch, probe, and close—each of which became a predictable module complete with scripted dialogue, uniform selling procedures, and expected time to complete.

Studies of factory and machine shop floors have shown that computer implementation decreased skills among blue-collar workers just as it had among service workers. The most studied technology to enter the machine shop floor was numerical control (NC). According to the labor and technology scholar Harley Shaiken, NC was not a new method of cutting metal "but a means of information processing and machine control. In fact, metal is cut in the same way as on a conventional machine, using the same types of drills and cutters. The difference is that an NC machine is controlled by pre-coded information while a conventional machine is guided by the machinist."[43] When managers assigned the task of preparing the precoded information to programmers or engineers rather than to machinists, machinists suffered a loss of skill, autonomy, and responsibility. No longer did they make decisions about how to cut metal; instead, they set up the machine as specified in the programmed instructions they received, tended the machine while it cut the metal, and removed the piece when the machine finished. Mainframes and minicomputers shortly replaced mechanical controls in NC systems, thus bringing about computer numerical control (CNC). In CNC systems, information from each machine in the factory or machine shop traveled to a central location, allowing the plant manager and his staff to closely monitor and control all operations. Such computerization further restricted the decision-making domain not only of machine operators but also of foremen, supervisors, and line managers.

Computer-engendered deskilling did not stop at service and blue-collar settings but extended into professional, managerial, and technical work as well. Managers and professionals from fields as diverse as social work and financial brokering found themselves in the crosshairs of computer automation, whose growing role usurped their skills, limited their decision-making control, and provided the means for quantitatively measuring their performance. Computerized information systems on Wall Street, for example, allowed upper management to closely monitor the performance of brokers, including how often they used their firm's "expert system" software programs.[44] These programs, which formalized and codified brokers' knowledge, posed a set of questions to the client and then, based on the client's responses, recommended specific company financial products for the client to buy. Particularly at risk were low-level managers across a wide range of industries whose primary task involved the collection and transmission of routine data, activities for which computers were well suited.

Common to most of these accounts was the recognition that the deterministic effects that flowed from the use of computerized systems were not the product of the technology itself. That is to say, materialistic determinism could not explain these results. Instead, deskilling theorists argued that using technology to separate the conception of work from its execution was the inevitable outcome of a dominant managerial ideology rooted in scientific management. In other words, particular technology choices were not attributable to the technology itself but to managers' goals and ideologies, as well as to managers' strategies for extracting the most productivity from their workforces.

Not everyone agreed that computers were deskilling workers. A number of scholars found that the introduction of computers into workplaces often meant that workers' skills improved. Computer work required higher analytical skills than did work before it, and workers, freed by computers from many manual tasks, found themselves with a broader range of responsibility. Rather than deskilling workers, these scholars claimed, computers "upskilled" workers.

The sociologist Robert Blauner provided early ammunition for those scholars who argued for upskilling. In a study of four blue-collar manufacturing settings (a print shop, a textile mill, an automobile factory, and a chemical processing plant), Blauner found that workers' loss of decision-making control did not parallel increasing computer automation of the work.[45] Rather, at higher levels of automation, workers regained a certain level of decision-making control and consequently felt less alienated than workers whose mechanized tasks featured lower levels of automation. Other studies similarly found that work at the extreme end of computer automation resembled craft production more than it did mechanized work. Compared to craft production, mechanized work featured a high division of labor and limited job scope for individual workers. Automation could not bear the blame for those changes because job scope was once again broad in the most highly automated forms of work.[46]

Upskilling was not limited to production environments. The management scholar Paul Adler reported the case of a large French bank that intended to deskill its lower-level workers, including tellers, with the introduction of an advanced computer system.[47] In accordance with the bank's plan, programmers readied entire sequences of operations for the computer to perform, and managers eliminated a number of lower-level jobs. Much to the bank's surprise, however, the remaining workers required considerable knowledge of computer processing systems and basic accounting to be proper stewards of the new system. The result was an aggregate upgrading of skill among the bank's workforce.

Overall, occupational data on manufacturing from 1976 through 1990 appeared to reflect a process of upskilling, largely achieved through occupational redistribution. In blue-collar work, the number of craft workers rose relative to the number of machine operators. More broadly, the share of clerical and administrative positions declined in comparison with sales, managerial, professional, and technical occupations.[48]

By following the logic of idealistic determinism, theories of deskilling and upskilling were able to explain consistency in technology choice without attributing causality to material features of the technology. These theories argued that ideas and beliefs shaped action and outcomes associated with the introduction of the computer into workplaces. The ideas and beliefs were born in external forces: in the case of computerization, deskilling theorists argued that scientific management drove a need for managerial control that strongly guided the ideas and beliefs of all managers. So ubiquitous was the paradigm of scientific management, deskilling theorists claimed, that all managers held the same ideas and beliefs and, acting on these ideas and beliefs, made the same technology choices, which resulted in the same outcomes in every workplace. Upskilling theorists changed this story only slightly: they argued that managers uniformly valued efficiency, which managers could more easily achieve by upskilling (and trimming), not deskilling, their workforce.

Theories of deskilling and upskilling fell prey to criticism from social constructivists, who, in finding empirical variation in managers' choices, contended that technology choices were not as universal as these theories claimed. In other words, not all managers shared the same set of ideas or beliefs, and for this reason managers must have acted in response to their local context, as social constructivism holds.

Although deskilling and upskilling theories fell short, idealistic determinism may still hold promise. At its core, idealistic determinism suggests that actors aim to accomplish specific goals in accordance with their ideas and beliefs, which guide their actions, and that these ideas and beliefs derive from external forces. The value of this perspective is that, unlike perspectives rooted in voluntarism, it allows prediction. That is to say, if one has a mapping between ideas or beliefs and choices, and another mapping between choices and outcomes, then predicting outcomes is simply a matter of knowing ideas and beliefs. Deskilling and upskilling theories failed not because they incorrectly presumed that ideas and beliefs shaped human action but because they presumed a universal mapping that allowed for only one set of ideas and beliefs. An idealistic deterministic perspective that allowed for multiple sets of ideas and beliefs and that constructed multiple mappings between these sets and technology choices might solve this problem and provide a viable path for explaining—and predicting—technology choices. We provide such an alternative by choosing occupations as the grounding for establishing multiple sets of ideas and beliefs.

An Occupational Perspective as an Idealistically Deterministic Alternative

To understand the appeal of an idealistically deterministic perspective on technology choices that is rooted in occupations, let us return briefly to Levy and Murnane. Levy and Murnane argued that, in the field of medical imaging, computers complemented the work of, but did not substitute for, cardiologists because the computer was not as good as a human in recognizing patterns in the computerized images of echocardiograms. The conclusion they drew from their (technologically deterministic) analysis was that cardiologists had no need to fear substitution as a result of computer implementation; this occupation was thus "safe" from automation and offshoring, two work solutions that, in recent times, have often accompanied the computerization of work.

One need only look at the case of diagnostic radiologists, who, like cardiologists, must read and interpret images created via computer technology, to recognize the problems inherent in this analysis. Today, doctors in U.S. hospitals routinely send radiology scans through digital connections to radiologists in countries such as India, who upload and examine them on their computer monitors, interpret them, and send their diagnoses back to the U.S. treatment team. In this manner, computers have had a significant impact on the work of radiologists by facilitating the substitution of U.S. doctors. Levy and Murnane's formulation could not have predicted this impact because, as in the case of echocardiograms, humans still interpret the image by means of pattern recognition and complex thought processes.

That the substitution of diagnostic radiologists was by other humans rather than by computers in no way alters the fact that substitution occurred and that the computer was instrumental in its occurrence.[49] In part, Levy and Murnane's formulation faltered because it evaluated the computer in isolation from the host of other technologies present in the workplace. By expanding the scope of inquiry to other technologies, in this case advanced information and communication technologies that facilitate the transmission of digital images, they might have better predicted technology use and outcomes.

More broadly, however, a host of occupational factors shapes who reads radiology scans and when and why they do so.[50] Hospitals send scans to India primarily during nighttime hours, when U.S. radiologists are largely unenthusiastic about being on call. During these hours, it is daytime in India, which means Indian radiologists, already paid much less than their American counterparts, are at their cheapest. This arrangement provides an economic incentive for offshoring the reading of scans. To protect their occupational standing, however, radiologists have worked through their professional body, the American College of Radiologists, issuing "technical standards" recommending that state-licensed and board-certified doctors sign scans. A handful of Indian radiologists meet this requirement, having gained their board certification in the United States before returning to India. In most cases, however, a U.S. doctor must review and sign off on the diagnosis that the Indian doctor provides. This requirement acts as a disincentive for the complete substitution of U.S. radiologists.[51]

Levy and Murnane's technologically deterministic explanation, derived from evaluations of what the work entails and what skills computers have versus humans, failed to accurately predict the role of advanced information and communication technologies in the context of medical imaging. A social constructivist perspective, on the other hand, would have held that hospitals would have chosen distinct work arrangements, yielding no consistent pattern of choices and outcomes. Yet that does not seem to be the case with teleradiology: hospitals across the United States, of all specialties and in all locations, are turning to offshoring solutions for nighttime radiological readings, with the radiology profession stepping in and dictating how hospitals can and ought to carry out offshoring.[52] The example of diagnostic radiologists thus points to problems in both technologically deterministic and social constructivist perspectives. More important, it points to the potential of a perspective rooted in occupations.

Occupations represent social groups with primarily economic bonds; they tend to have a shared normative order and a clearly defined set of knowledge and skills to be applied to a particular set of problems or tasks. Occupations are likely to employ special-purpose technologies tailored to their needs, such as applications for tracking accounts receivable and payable in accounting, simulation software for modeling performance in engineering, and image-rendering tools for altering photographs, copy, and artwork in graphic design. Although members of a given occupation often have similar educational and training backgrounds and work in the context of the same legal and regulatory framework, they work across a variety of organizations and settings.[53] Institutionally, occupations often shape the role that new technologies play by setting policies about them, just as the radiologists did in the context of transmitting digital medical images across large distances. Consequently, we often see homogeneity across firms in the outcomes of a technology within an occupation.

For these reasons, an occupational perspective on technology choices would allow for differences within individual settings (across occupations) while granting the possibility of similarities across settings (within occupations). That is to say, an occupational perspective would look for explanatory factors in neither the technology itself nor the realm of individual organizations but in the larger occupational environment. Such factors might include the history of professionalization within the occupation, the legal and regulatory framework that guides the occupation, or features of the product or service associated with the occupation.

The appeal of an occupational perspective is that it permits an account of why technology choices are similar across many organizations, yet not universal. It would also allow some amount of prediction of managers' and workers' technology choices: on the one hand, an occupational perspective would avoid the risk of overly broad, and hence inaccurate, generalizations; on the other hand, by permitting less sweeping generalizations, it would allow us to do better than resign ourselves to the conclusion that everything depends on local, situated factors specific to a given setting. Turning an occupational lens on questions of technology choices, as opposed to employing other potential perspectives such as sociomateriality and critical realism, could potentially yield useful generalizations that would help predict not only changes in knowledge, skills, and abilities within a given occupation but also the potential and often dramatic changes to the composition and global distribution of an occupation's workforce. Such information could guide everything from individual career choices to strategic management decisions among firms to government policy formation.

To specify the benefits of an occupational perspective and to elaborate what such a perspective might look like, we offer here a study of engineers in product design. Engineering affords a particularly useful starting point for this inquiry: engineers are representative of professional and technical workers, whose share of the U.S. workforce has grown considerably in the years since the introduction of computers in the workplace, up from 8 percent in 1950 to 17 percent by the end of the century.[54] Engineers work on products and systems that are often considered critical for American productivity, security, and technical supremacy. Educators, government officials, business writers, and policy pundits have engaged in considerable recent debate about whether or not America is producing enough engineers, especially in comparison with China and India.[55] A key element in these discussions is the question of how to make engineering programs in universities more attractive to U.S. college students, including women and minorities, and, more generally, how to promote the study of science and math in K–12 education so that students might be prepared for engineering studies at the university level.[56] In sum, engineers constitute a critical segment of the U.S. workforce, a fact that underscores the importance of understanding their technology choices.

From a pragmatic standpoint, engineering also provides a good place to begin an investigation of technology choices because no matter what their specialty or product, nearly all engineers engage in the fundamental tasks of design and analysis. In addition, engineers across occupations have available to them a multitude of sophisticated computational, logic, and graphical computer applications tailored to the tasks of design and analysis. These similarities in work and technology permit meaningful comparisons across engineering occupations.

At the same time, engineering occupations differ in other ways that should allow us to tease out important explanatory factors. For example, different engineering occupations were established in different periods in history and vary in the degree to which older knowledge in the field is relevant for modern-day practitioners. One occupation we studied, structural engineering, falls under the larger rubric of civil engineering, which is the second oldest engineering specialty. Civil engineering was founded in the 1740s in France and the 1770s in the United States; the subspecialty of structural engineering is well known for its continued reliance on basic principles from physics and material science that have remained unchanged for centuries. Structural engineers spend their careers amassing this knowledge, adding to an ever larger personal store of highly relevant work knowledge. By contrast, hardware engineering, the second occupation we studied, has its roots in electrical engineering, a field that was established more than a hundred years after civil engineering. Hardware engineers conduct their work in a world where knowledge, like products, becomes obsolete very quickly, forcing practitioners to constantly gain new knowledge so that they may remain useful contributors. This difference in the rate of knowledge change within an occupation, we found, influences how and why engineers employ advanced computer and information systems, and to what effect.

Similarly, engineering occupations differ in how the tasks of design and analysis are distributed among individuals within a single firm. In structural and hardware engineering, the person who designs a feature or an element of a larger product also typically engages in significant analysis of his creation. In automotive engineering, the third occupation we studied, managers organize design and analysis tasks such that one engineer designs a part or subsystem and one or more other engineers analyze it. We found that these differences in the division of labor also have a hand in shaping technology choices.

In fact, current efforts to divide the labor of automotive engineers bear a striking resemblance to efforts a century earlier to rationalize the work of automotive production workers. This similarity raises the distinct possibility that advanced information and communication technologies may be poised to become the assembly line of engineering work in automotive companies, designed to streamline and speed up the completion of small tasks spread across many workers. There is a certain irony in the possibility that the routinization of thought and action that engineers facilitated a hundred years ago when making technology choices for factory floor workers may now be coming home to roost in a place long considered safe from standardization: the office setting of the creative, highly educated, white-collar technical worker.

2 Automobiles, Buildings, and Chips: Product Design Work across Three Occupations

What does the work of product design look like? Scholars tell us that product design is a messy, social affair, a tangled web of interactions and negotiations, in stark contrast to the highly linear and rational design process that engineering textbooks typically portray.[1] Our research confirmed that engineering design involves a great deal of engagement among a host of individuals. Indeed, the path from concept to final product among the firms we studied was never straightforward: a multitude of modifications to the design marked progress, with each modification necessitating looping back to earlier points in the process to repeat tasks. This picture of repeated cycles of design work was strikingly similar across the three occupations we studied, namely, automotive engineering (automobile design), structural engineering (building design), and hardware engineering (digital chip design).

In this chapter we lay out the specifics of design and analysis in the context of automobiles, buildings, and chips so that we might take a deep look at the work these engineers perform. Our purpose in examining the specifics of product design work is twofold. First, we want to provide an understanding of the work activities in which the engineers we studied engaged so that our later discussions of how and why they made technology choices are all that more clear. Second, we want to establish that, because the three occupations share the fundamentals of design and analysis, we cannot explain differences in the engineers' technology choices based on differences in the kind of work that the engineers performed. For the sake of clear exposition and ready understanding, we discuss these design activities in their stylized, linear order, but to be true to the actual process, we also note how and why the process went amuck, with work cycling back to earlier steps and activities repeating themselves.

The automotive engineers we observed designed and analyzed the various parts (e.g., brakes, doors, seats), systems (e.g., ventilation, suspension, powertrain), and structures (e.g., frame, chassis, bumpers) that constituted vehicles. Whereas IAC designed most vehicle systems and structures, it hired suppliers to initially design many parts. IAC engineers therefore analyzed parts, systems, and structures to ensure conformance to guidelines and tolerances. They then assembled the parts, systems, and structures to create vehicles—cars or trucks—and analyzed the vehicles against broader performance metrics (often set by government regulation), including how well the vehicle fared in a collision, how much fuel the vehicle consumed, and how quiet the vehicle was on the road. The engineers conducted virtual as well as physical tests. Using CAD software, the engineers created detailed blueprints of vehicle components that served as input to sophisticated computer simulations for virtual (mathematical) tests. Other engineers built physical parts, and then assembled the physical parts into actual cars and trucks to conduct physical (real) tests at the company's proving grounds.

The structural engineers we studied specified the materials, shapes, and sizes of the beams, braces, piers, and other structural elements that transferred building loads to the ground to prevent buildings from collapsing. Hired by architects to work on specific projects, structural engineers created documents that illustrated alternative designs for architects and building owners. After the architect and owner on a particular project chose a conceptual design from among the alternatives they received, the structural engineers formalized and analyzed the building design to ensure that it met the performance requirements specified in state building codes. They also produced detailed drawings and calculations to help contractors estimate the project's cost and to guide the construction of the building. The structural engineers presented their final designs to county review boards, which issued building permits after approving designs.

The hardware engineers we observed crafted the logic of microprocessor cores, buses, transceivers, and other digital chip components. Working on project teams for each version of the microprocessor, they created representations of these components by writing code in high-level programming languages. The code specified how each component would handle the instructions that the component received and how the component would interact with other components. The engineers also wrote test programs to analyze their components and verify performance. Although at first glance one might assume that the hardware engineers were software programmers, their work required proficiency in the logic of information flows and a keen understanding of chip design.

As these overviews make clear, design and analysis activities are at the core of work in each of these three occupations. In everyday practice, design and analysis entail a variety of activities, such as performing calculations, making judgments about methods, creating mathematical models, drawing, sketching, solving problems, running simulations, interpreting test results, negotiating changes, and exchanging technical information with colleagues and clients. Design and analysis are at once social and technical, collaborative and independent, art and science, estimation and precision.

Although the particulars of design and analysis varied by occupation, the work process was similar across the engineers that we studied. In each case, engineering work began with an idea for a new product. The job of the engineer was not to come up with that idea but to figure out how to bring it to fruition. The first step toward achieving that goal was to derive a general solution that laid out the main features of the design, the unique attributes that would do most of the work in getting from concept to reality. With this general solution in hand, engineers would divide up the detailed design work among themselves. The division first split the product into components, then assigned engineers to the components. In the case of autos, these components might have been subsystems, such as the powertrain, the chassis, or the body, which were then further divided into constituent components, down to the minute level of handles, hooks, and supports. For buildings, the components were often the individual floors of a multistory building. For hardware engineers, the components consisted of buses, caches, accelerators, transceivers, and the like.

Engineers working on each component would hammer out its details. In structural and hardware engineering, these same engineers tested and analyzed the performance of the components that they designed; in automotive engineering, other engineers in the group did these tasks. When testing and analysis were satisfactorily completed, the engineers would integrate their separate components back into a whole. The assembled whole was then ready for further testing and analysis, which may or may not have been done by this same group of engineers. Analysis of the whole frequently pointed to failure of one kind or another in one or more components, which would require some subset of the engineers to return to an earlier step in the process and begin work anew from there. Hence, across all three occupations, the engineering work process entailed repeated cycles of design and analysis.

Beginning with an Idea

All engineering design in the firms we studied began with a rough idea of what the product should do and how it should look. None of the engineers we observed came up with this idea; rather, they entered the design process after someone else had formulated the idea.

In automotive design, the process began with marketing. Market researchers at IAC surveyed automobile drivers, passengers, owners, and others about what they were looking for in vehicles and how IAC vehicles compared with competitors' vehicles. Members of the marketing team also worked closely with dealerships, hoping to learn what needs and desires consumers professed in showrooms and what opinions they had after viewing and test driving IAC vehicles. As Marie, a senior marketing manager commented, "Dealers have a lot of power in the automotive world because they tell us what they can and cannot sell. They know it because they are on the front line. So we listen."

After the marketing team had assembled a rough sketch of the kinds of vehicles that would be attractive to customers, it would work with senior executives to determine what alterations would align with IAC's product strategy (at the time of our observations, the strategy was to increase fuel efficiency) and to assess how certain changes to features, performance, size, and styling would affect a vehicle's brand image. The executives would then determine appropriate pricing for a vehicle and the date by which IAC should release the vehicle to the market to gain the greatest possible market share.

The senior executive teams that made these decisions consisted mostly of heads of business units. Typically a lead vehicle architect was also involved in this decision making. The lead vehicle architect was responsible for ensuring that IAC could eventually build vehicles that, at this point, existed only in rough specifications and sketches. To assess the engineering feasibility of a design, the lead vehicle architect would give these rough specifications and sketches to engineers at the company's Advanced Vehicle Development Center, or AVDC.

The AVDC engineers normally began their assessment by gathering computer simulation models of the current generation of the vehicle, then would use these simulation models to make the changes the senior executives had proposed. They also contacted R&D engineers to learn whether new materials or new technologies were mature enough to include in the proposed vehicle's design. The AVDC engineers would run rough analyses of the design to assess whether it achieved desired performance (e.g., fuel economy, safety rating).

After the AVDC engineers had completed these preliminary analyses, the senior executives would sign off on the overall design of the vehicle, which at that point was little more than a list of features the vehicle would have and the performance levels it would achieve. The lead vehicle architect would then outline a detailed "vehicle development process" that enumerated deadlines for each stage of design and analysis. Typically, the lead vehicle architect would work with engineers in AVDC for nearly a year to translate the list into actual specifications for the vehicle.

In building design, the process began when the owner of a property hired an architect to design a building. The architect would develop the design idea for the building in conjunction with the owner by asking what kinds of human action and interaction each part of the space should facilitate. With the design idea completed, the architect would explain to the structural engineering firm the project's goals, budget, and other constraints.

The job of the structural engineers was to embody the architect's idea in a structural solution that met the project's goals and constraints. For example, an architect's vision of a building might feature an open, airy lobby devoid of columns. The structural engineers first had to determine how to achieve such a lobby. Perhaps the answer lay in a king post truss that could deliver the weight of the building to the walls absent support columns in the lobby. The engineers would arrive at such a conclusion only after a lengthy social process of negotiation with the architect, and perhaps the building's owner. After initially meeting with the architect, the structural engineers would return to the office. Senior engineers—typically principals in the firm—would then develop several feasible strategies (e.g., using steel versus concrete). They would consider two, three, or four alternatives and sketch, by hand, their ideas. After several days of this activity, the principals would discuss their most promising ideas with the architect.

Eventually, after discussions with the architect, the principals would boil the alternatives down to generally no more than two options. At this point they would hand off the project to a single senior engineer to do rough, preliminary calculations. The engineer would do these calculations manually (i.e., with a calculator, a sheet of paper, and a pencil) because he did not yet have a well-defined model of the building. His ideas were still creative, not analytic. As Roger, president of Seismic Specialists, explained, "We can't analyze yet or every building would be a research project." The point of the calculations, then, was to help the principals think through broad issues such as constructability and compatibility.

With rough calculations in hand, the principals would sit down with the architect and the owner to discuss the merits of each option. If the engineers had done a good job in understanding the building's intended use, the owner's desires, and the architect's vision, one of these two options was likely to be agreeable to all parties. It was not unusual, however, for the architect to rebel against a structural design because he favored a different material (e.g., wood over steel) or because he was unhappy with how the design would affect the building's aesthetics. In such cases the engineers would have to cycle back through the process one or more times.

In chip design, as in automotive design, the marketing team, based on feedback from the firm's customers, determined what the chip should do. The marketing team formalized its idea in a set of specifications, which it handed off to the chief architect, whose job was to determine what functionality the chip required to achieve the marketing team's vision, and how to deliver that functionality. The chief architect, acting largely alone, carried out the process of determining how to convert the idea into design.

As a sign of the importance of this role, chief architects were the highest status engineers in the firms we studied. At Configurable Solutions, the chief architect was the only nonmanager to inhabit, without a roommate, a glass-walled office. On the walls of his office hung six patent plaques; no other engineer we observed held more than two. When engineers went to speak to the chief architect, they did not venture further inside his office than the doorway. A raised eyebrow, a slight frown, or a bemused smile as he listened to their questions suggested that he did not consider them his equals. He did not, for example, explore with them for hours the contours of their problems or accompany them to their desks to hash out the details, as we observed many engineers doing when approached for help by a colleague. Instead, he often just reminded them about what the instruction set architecture manual, which he wrote, specified. As Eric noted,

> All of his [the chief architect's] design is done through email. He says do this and we [the hardware engineers] are his tools. This [*Eric picks up a copy of their instruction set architecture manual, which on another occasion Eric referred to as their "Bible"*] is what should end up in here [*he points on the screen to his code*].

The chief architect had two primary decisions to make. The first decision was to determine what blocks were needed to achieve the marketing team's desired functionality for the chip. (Blocks are groups of transistors that engineers arrange and program to perform a given function. Examples include cache blocks, whose function is to store memory, and ALU blocks, or arithmetic logic units, whose function is to compute numbers and perform basic logic operations.) The second decision was where to place each block on the chip. Combined, the blocks and their specific arrangement on the chip constituted the architecture of the chip. In determining the layout, the chief architect had to weigh the trade-offs among a number of competing factors, including cost, power, performance, quality, and ease of use.

Dividing Up the Work

After the lead actors in the firms we observed had agreed to a general solution, the detailed solutions had to spell out the exact materials, specifications, functions, shapes, and sizes of all components in the design. Because that job was typically too big for any single engineer to handle, engineers would divide up design work at this juncture across individuals, teams, or entire departments.

In automotive engineering, the AVDC engineers would hand off the design to product development engineers to begin design and analysis of the vehicle's parts, systems, and structures. Managers would divide product development work by functions that corresponded roughly to vehicle systems (such as the powertrain or chassis), and would further divide work into specialties of design and analysis. That is to say, the managers would separate out one group for design (we will call them design engineers) and one group for analysis (we will call them analysis engineers). Within the design group, managers would further divide work by assigning groups to major sections for each vehicle. Within these section subgroups, managers would charge individual engineers with the design of up to thirty distinct parts each. The assignment of individuals to parts ended the division of labor within the design group. Within the analysis group, the engineers would most often analyze the performance of the entire (integrated) vehicle. Managers would divide analysis engineers according to sets of specific performance concerns, such as noise and vibration, aero/thermal dynamics, and safety and crashworthiness.

In structural engineering, the division of labor was much simpler than in automotive engineering. Principals would hand the work of providing detailed solutions to a single senior engineer and his or her team, which typically consisted of one mid-career engineer and two to three junior engineers. Teams would divide up their work by assigning building sections or floors to specific engineers. For example, at Tall Steel, a team of engineers completed the design of a five-story luxury hotel. Sam, a senior engineer and one of four founders of the firm, was in charge of the project. Working with him were two junior engineers, Darren and Beverly. Darren and Beverly split up the hotel's five floors between them. The laws of physics dictated that each floor had to bear the weight of all the floors above it. One might mistakenly conclude from this fact, as we did, that the engineers had to complete the design of all uppermost floors before they could design any lower floor, but this was not the case. Darren did not wait for Beverly to complete her floors before designing his, which lay below hers. Instead, he used estimates of the floor loads above his floors so that he and Beverly could work independently.

The division of labor in hardware engineering was just as simple as that in structural engineering. A group of hardware engineers would implement the architecture for a new microprocessor based on the chief architect's high-level description of blocks and the blocks' location. Managers would assign engineering subgroups to different blocks; within these subgroups, individual engineers were responsible for designing specific block components. Each engineer would map out his design for his component via a flowchart or diagram that represented elements with rectangles and other shapes, all connected via arrows. Because these diagrams represented the design, they stayed unchanged for months (like the one on Eric's whiteboard) and served as big-picture touchstones for the engineers whenever they had to puzzle out how to connect two components or debug a failure.

Hammering Out the Details

With product components distributed across multiple engineers, the design process moved into a stage in which engineers needed to hammer out component details. In automotive engineering, design engineers worked closely with parts suppliers to see if off-the-shelf components might fit the vehicle's needs. Design engineers also worked closely with manufacturing engineers and technicians, who advised them on which materials were best to use and confirmed that parts were buildable. With this input, each design engineer would design his or her part using CAD software, assigning the part's coordinates on a three-dimensional grid to indicate where in the dimensional, spatial system of the vehicle the part would be located. The design engineer was the only person at IAC who had permission to alter the master CAD file for the parts that he or she "owned." Design engineers stored these master CAD files in a database called DataExchange so that other engineers might access (but not alter) the files.

Design engineers would often order physical tests to determine their parts' robustness, to verify that their parts met specifications, or to make sure that their parts operated properly. For example, a design engineer might order a door slam test in which automated equipment would slam the door shut 10,000 times to test its endurance. The results of the tests helped design engineers determine their parts' specifications in terms of materials, shape, and the like.

Design engineers also benefited from tests that analysis engineers who were assigned to parts groups performed for them. Although most analysis engineers analyzed full vehicles, some analyzed parts, systems, or structures to aid design engineers in developing their specifications early in the design process. For example, Melissa, an analysis engineer in the body structures group, was responsible for ensuring that the upper body structures could support the weight of the vehicle's roof and doors. During one of our observations, Melissa was analyzing sheet metal parts in the upper region of the vehicle's frame. Through her analysis, Melissa came to realize that the design engineers responsible for the sunroof harness and window wiring attachments had specified that holes be drilled into pillars to mount their parts. Melissa's analysis determined that the requested holes dramatically weakened the strength of the pillars, thus compromising the integrity of the structure. To fix this problem, Melissa had to meet with the design engineers individually to ask whether they could either remove the holes or reduce the holes' diameters.

One of the design engineers that Melissa met with was Gabe. Gabe was responsible for the design of sunroof components, and he had specified in his master CAD file that a hole be placed through a pillar for an electrical conduit that would power the sunroof. In advance of their meeting, Melissa sent Gabe a PowerPoint slide deck with screenshots taken from her simulation model of the integrated vehicle structure. At the beginning of the meeting, Melissa asked Gabe to open the slide deck so that she could show him the problem with the holes he requested. Melissa briefly explained the problem and proposed how she wanted Gabe to change his design. As Melissa answered Gabe's questions about such matters as the location and number of holes permitted and whether or not he could use nuts for attachments, he began to understand how he needed to change his design to address the integrity concerns that Melissa had identified in her analysis.

In this manner, design and analysis engineers at IAC hammered out the details of component design in slow cycles of part specifications, testing, and respecification. Although the rough specifications for parts that AVDC gave to design engineers directed initial design decisions, the detailed feedback that analysis engineers provided shaped the ultimate form and location of parts. The fact that design and analysis engineers sat in close physical proximity (often on the same floor of a large building or in adjacent buildings on the same campus) made frequent face-to-face meetings possible.

In structural engineering, the team began its work with the single solution for the building, which specified the building's size, shape, mass, grid, and framing type. Starting from this base, the team's first step was to create a general analytical model. Depending on the complexity of the project, the team could create the model manually (meaning with calculator, pencil, and paper) or on the computer. To create the model, the engineers used approximations and estimates based on solutions from prior projects or intuition they built up from experience. When completed, the analytical model consisted of a project binder full of calculations and a preliminary sizing for all the beams (their sizes, weights, shapes, and types), the framing, and the foundation. The purpose of the model was to facilitate estimation of the building's cost.

The next step in structural engineering was to move from estimates to precise values, namely, to develop equations whose solutions would specify all the details in the analytical model, right down to the number and size of bolts per beam. Because idiosyncrasies in the form of special features (such as odd roofs or unique overhangs) marked each building, the engineers often struggled to make the right assumptions, develop the models, and do the necessary calculations to solve these equations. Because the calculations served as justifications for decisions concerning the size and shape of all components, the engineers had to derive explicit, defendable answers.

The primary analysis that engineers completed at this juncture was the gravity analysis, which determined what size and shape components needed to be to handle the downward force, or gravity load, of all components above them without breaking, cracking, slipping, shearing, or otherwise failing. Because engineers had to conduct gravity analyses on each floor, and because idiosyncratic building features defied standard solutions and thus demanded individual attention, the engineers we observed spent a great deal of their time working on projects at this stage. For junior engineers, whose engagement with architects and building owners was limited, this work consumed the bulk of their time.

Senior engineers checked in with junior engineers to make sure that work in this stage was progressing smoothly. In the case of the luxury hotel that Tall Steel designed, Sam regularly met with Darren and Beverly in a small conference room, where he would lay out drawings on a circular table, around which they would gather. In meetings that could last for hours, he would review their work and help them sort through issues on their floors. The meetings served as opportunities not just for project review and problem resolution but for teaching and learning. For example, in discussing why certain design solutions were good and others bad, Sam schooled his young colleagues in matters such as why their solutions could not require carpenters and ironmakers to be on site at the same time (because these two groups did not get along) and how to design with the size of a workman's fist in mind (to be sure that a workman had room, for example, to rotate bolts in narrow spaces). Project leaders at Seismic Specialists and Tilt-Up Designs held similar meetings with their teams.

In hardware engineering, the engineers would hammer out the details by converting the design of their components, as laid out in diagrams on their whiteboards, to working models. Each model was in the form of a register-transfer-level (RTL) abstraction. RTL is a way to model registers and the logical operations performed on the signals between registers. Registers are the components that store bits of information, to which other registers can send ("write") or receive ("read") information. An example of a logical operation performed on the signals between registers is, "If the clock signal goes to 1, then store in Register A the contents of Register B." The engineers used a hardware description language, such as Verilog or VHDL, to create RTL when implementing their components. Viewed on the computer screen, Verilog code looks a lot like any programming language, with if-then-else statements, do-loops, comment lines, and equations that assigned values to variables. What makes Verilog and other hardware description languages more than software programming languages is their representation of clock time, an essential feature when modeling events and the occurrence of events within a microprocessor.

A good bit of the engineers' work in writing RTL code involved figuring out how to implement their design. In the example below, Doug, a hardware engineer at Programmable Devices, pointed to a schematic of the chip design on his monitor as he explained to the observer the problem facing him in the design of his component. Central to the new design was a BRAM (memory cache) block, located just below the multi-gigabit transceiver (MGT), a component that converted parallel data to serial bits. Accompanying the BRAM and the MGT were new clocks, whose paths Doug needed to route optimally:

> The question we are facing is how to handle the clocks. On the old version of the microprocessor, we had two high-quality reference clocks.... We wanted to do better. So what I've been looking at is how these things actually get routed around, and kind of plan it such that the clock routing will use good routing to preserve the quality of the clock. So eventually the reference clock, BREF clock, will get in to the MGT. The MGT itself creates an output clock, which is called TX OUT clock. And this output clock may need to go to a DCM [digital clock manager] or to a BUFG [a buffer used to relay clock signals], and at that point it gets buffered or new clocks get derived and those clocks will be distributed to all the MGTs, to all the 10-gigabit transceivers. And what I'm looking at is the routing from TX OUT clock to all these different places, well, actually, to BUFG and the DCM, and what we would need to do to make that a good route.

Having solved their technical problems and coded their components in RTL, Doug and other hardware engineers would next verify that their code worked properly by testing it. Testing required writing a test program, which the engineer would do himself. Tests could either be directed (the engineer knew which instructions he wanted to test and in what order) or random (the engineer wanted to see whether errors emerged with sequences of instructions he may never have thought to run). Many engineers chose to write their tests in a hardware verification language called Vera. Coding tests in Vera looks a lot like coding RTL models in Verilog: both tasks involve the hardware engineer sitting in front of a computer monitor writing lines of programming code, checking syntax, borrowing sections of code from existing files, and puzzling out what needs to be done and how to make it happen. Writing and running tests of the component that they themselves designed meant that hardware engineers were fully integrated into both design and analysis work.

Integrating the Components, Testing the Whole

Engineers know better than to wait until product components are integrated into a single whole to worry about how the components will function together. Instead, engineers consider surrounding components when designing their own component. When the structural engineers we observed made educated guesses about the load of the floors above their floor, even as their peers were still crafting those designs, they exemplified this concern about surrounding components. Similarly, the automotive engineers and hardware engineers, whose products were often new models of existing products (e.g., the annual update of an automobile model or the latest version of a microprocessor core), routinely used previous instantiations of surrounding components as reasonable stand-ins for the not-yet-designed new instantiations. Of course, estimates are exactly that, and when the engineers finally assembled all the components into a whole and began testing it, typically many problems would surface. Thus began multiple rounds of testing, analysis, and subsequent adjustments to the design of components.

In automotive engineering, analysis engineers outside the parts groups conducted the primary testing of the whole. Using the design engineers' master files in DataExchange, these analysis engineers assembled parts into fully integrated models that contained all the parts of a vehicle (generally more than 30,000 parts). The engineers tested the performance of the assembled vehicle in terms of crashworthiness, noise and vibration, aerodynamics, heat transfer, and cooling, among other areas, depending on each engineer's subspecialty group.

To complete these performance tests, analysis engineers used computer-aided engineering (CAE) computer applications. These CAE applications differ from CAD technologies in that they facilitate *analysis* rather than *design*. That is, analysis engineers use the complex mathematical equations that underlie CAE applications to determine the state (mechanical, thermal, magnetic, and so forth) of components or entire vehicles. The analysis engineers we observed typically utilized finite element analysis (FEA) or computational fluid dynamics (CFD) analysis techniques in the CAE application to test performance.

Both FEA and CFD techniques involve the creation of a mesh. A mesh is a grid of triangular or rectangular small shapes, called elements, superimposed over the surface of a component or vehicle. Although the mesh has a graphical form that engineers can visualize on a computer screen via a realistic 3-D image, a more important feature of the mesh is that it also has a numerical representation, in matrix form, within the computer application that created it. In this matrix form, which records the spatial location of each of the nodes, or corners, of every mesh element, the mesh is fit for computational analysis. That is to say, engineers could alter the numbers in the matrix through a series of computations that simulated some event, such as a car crash. The resulting new matrix at the end of the computations would represent the state of the component (for example, the spatial coordinates of all the nodes of all the mesh elements) after the event. A changed coordinate for a node, in the case of a simulated crash, represented plastic deformation of the component, as when a bumper became dented. On the screen, what appeared graphically was the new shape of the component. Creating good meshes was therefore essential for getting good simulation results.

Using a preprocessor like Hypermesh or Easi-Crash, both commercial software applications, engineers would create a mesh for each component and then assemble the components into systems or a vehicle. The next step was to populate a model with a series of equations representing boundary conditions for the mesh. Examples of these conditions included the load that would be placed on the structure and the velocity at which the vehicle would be traveling. In addition to a set of boundary conditions, each test required that engineers place sensors—called accelerometers—in specific locations in the virtual vehicle to record the vehicle's movement. Specifying the boundary conditions and placing the sensors were part of a task called "model setup," which engineers also completed on the preprocessor.

Engineers submitted their setup models to a "solver," which ran on a series of UNIX workstations linked in parallel or on an IAC supercomputer to solve the model's equations.[2] When the solver finished, analysis engineers used commercial postprocessing software to obtain pertinent performance results.

Model building required judgment and skills that analysis engineers gained through experience. One skill was to know which parts had to be included in a model and which could be excluded. Excluding parts simplified calculations and speeded up simulation runs, but missing parts, if actually needed, rendered simulation results invalid. Gretchen, a safety and crashworthiness (S&C) engineer, explained how she gained this skill:

> Building a model is both art and science. For me, the art comes from knowing what to include and what to leave out. You just learn this slowly over time by seeing what other people do and from trial and error. You look and learn and you ask people why they didn't include the plenum[3] or something and they tell you and you make a note in your mind that you don't need to include the plenum. It's not like you can write that all down because for each loadcase[4] you're doing you have to know how the energies move through the vehicle. If you know that, you can start to get a sense for what to include and what to leave out.

Whereas the "artistic" part of model building involved deciding which parts to safely exclude, the "scientific" part involved deciding what to do when simulation results did not reflect desired vehicle performance. In that case, the engineer had to make a design change, run the model again, and see whether performance improved. If it did not, she had to try a different change. This constant iteration of design changes was methodical and, at times, tedious as engineers repeatedly revisited the preprocessor, the solver, and the postprocessor. Recording each iteration's results in handwritten tables, electronic spreadsheets, or Post-it notes stuck in sequence across their desk, engineers tracked whether or not changes improved performance. When they had determined the optimal solution, or maybe simply a satisfactory one, they would try to persuade the design engineer who owned the part to make the relevant change.

Arvind, the analysis engineer from Bangalore whom we met at the beginning of this book, illustrated this iteration in design and analysis. Arvind was responsible for conducting rear-impact analyses for one of IAC's marquee sports cars. The initial tests he had run indicated that a rear impact would cause the fuel tank to crack and leak fuel. After several weeks' work, Arvind had determined that the solution that would result in the least design change would be to add dimples to the steel rear rails of the chassis.

These dimples, which S&C engineers called "crush initiators," made the chassis weaker in the rear of the vehicle. The goal of weakening the chassis was to have it bend consistently and fully, turning kinetic energy into heat that would dissipate with each accordion-like fold of the rear chassis. More energy dissipation meant less jarring, a safer crash event, and no rupture of the fuel tank.

To determine the size and location of the crush initiators, Arvind had created forty-eight different design alternatives and was testing each one. Although crush initiators represented the minimal disruption to the current design, they were still a major design change. For this reason, Arvind had met several times with Bilaji, the design engineer who owned the rear rails. Bilaji said that his manager was likely to resist the change, and he asked Arvind to document in detail how each design change was better than the one that came before it. Bilaji gave Arvind a PowerPoint deck and asked him to place the test results from each of the forty-eight design iterations in a table on a slide. Bilaji also asked Arvind to include pictures of the various tested designs. Figure 2.1 displays the picture of the crush initiator design that Arvind tested in alternative no. 43, discussed in our excerpted field notes below that document Arvind's meticulous work.[5]

10:50 a.m.

Arvind opens a plot in the postprocessor that displays the dynamic intrusion of the rear rail for iteration no. 43. The plot shows a line which represents the deformation

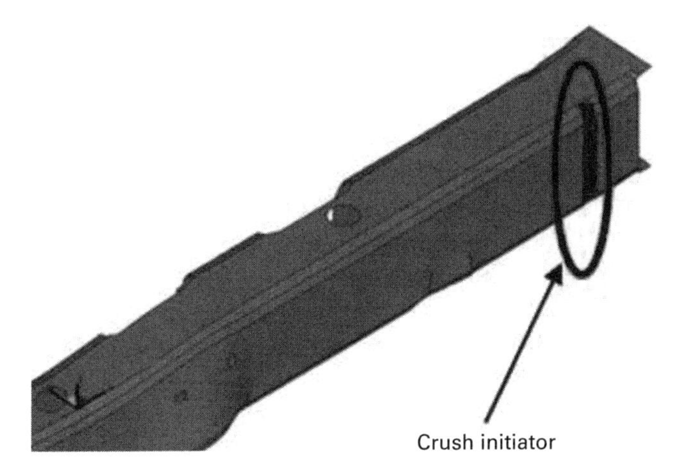

Figure 2.1
Arvind's Image of Crush Initiator Design no. 43.

of the barrier [which struck the vehicle from behind] and a line which represents the deformation of the rail.

Arvind: [*to the observer*] I pick one of the nodes on the barrier and one on the rocker [a suspension component to transfer wheel loads] side so when it crushes the difference between both nodes it give how much the barrier is intruding into the vehicle. By subtracting both the curves I get a curve of what the barrier intrusion looks like.

11:00 a.m.

Arvind subtracts the curves for the right-hand and left-hand sides of the rail and puts the numbers into the table in the PowerPoint deck that Bilaji sent. He generates a new plot in the postprocessor for iteration no. 44 and then manually copies numbers from the plot into the table. He also types them into a calculator on his UNIX desktop to calculate the % crush, which he enters into the table. Arvind repeats these steps as he moves across the table, filling in columns and rows.

11:20 a.m.

On the preprocessor file of the rear rails, Arvind selects several elements and then opens a window to calculate mass. He puts the resulting number in the table.

Arvind: [*looking at the result*] Those two little initiators result in 80 grams of savings. That is a significant reduction.

Arvind opens an image of the fuel tank in the preprocessor and calculates the plastic strain. He takes a screen shot of the image and pastes it into the PowerPoint slide. I ask him why it is important to add this image of plastic strain into the presentation. He explains that Bilaji's main interest is in understanding the fuel tank and how the rails need to be redesigned so that they soften the crush, so this image is really the most important component for him.

As Arvind later explained to the observer over tea, this tedious but important work to determine the best design for the rear rails took its toll on his personal life. To meet the deadline, he often stayed at work until 8:30 p.m., or two and a half hours after his shift ended, much to the dismay of his family. The work also took its toll on his storage space on IAC's servers: Arvind filled his allotted space, and consequently could not work on other projects.

In structural engineering, the primary testing required of the whole is generally twofold: lateral analysis and foundation analysis. Through lateral analysis, engineers examine how lateral forces, or forces perpendicular to a building's walls, will affect a structure. Examples of lateral forces included seismic forces in California or wind forces in Florida. If the analysis indicates structural failure, the engineers modify components until the structure can successfully withstand the type and size of lateral forces that characterize its geographic location. Lateral analyses were the most computationally intricate of all the analyses that the structural engineers we observed conducted, often calling for FEA models using technologies similar to those in automotive engineering but tailored for structural engineering. Like automotive engineers, structural engineers often employed multiple technologies when analyzing FEA models. Because engineers had to use the results from gravity analyses as input for their models, this step further required that they gather prior testing data from their colleagues for the many components that constituted the whole.

An observation of Darren exemplified the types of information and technologies that engineers employed in lateral analysis. Sam had decided that they should replace a particular building's V-shaped frames with X-shaped frames because its FEA models were failing. To carry out this decision, Darren began in Risa (a commercial software FEA application), where he changed all his V-frames to X-frames. He then compared Beverly's gravity analysis results, stored in an Excel spreadsheet, with numbers that Sam had given him on a handwritten sheet. In his own Excel spreadsheet, Darren replaced numbers in a table with numbers from Sam's sheet. He then opened some CAD drawings, locating on his screen several columns of interest. Because he could not have CAD and Risa open at the same time on his computer (the computer would crash if he tried), he copied down from CAD to a piece of scrap paper numbers for the columns, and then closed CAD and opened Risa, where he entered the numbers into a table. Next, he opened the *Manual of Steel Construction* and typed a formula from it into a table in Risa: 1.2D + 1.0E + 0.5L.[6] He plotted his results. Inspecting the plot, he announced, "Red arrows are bad. I have one." He remarked, "One option is to do a redesign. The W10×33 is the only one that failed.[7] Everything else is really low.... I wonder if the steel grade is wrong?" He checked the *Manual of Steel Construction* again. "Yep," he said. "It's 50 for yield stress; not the 36 I had entered." He changed the number for the steel grade, reran his model, and saved it in Risa.

Structural engineers typically conduct a second analysis, a foundation analysis, to determine how to dissipate a building's combined gravity and lateral loads into the ground. Typical solutions include introducing foundations, piers, piles, and other elements having a direct connection to the soil. Foundation analyses do not entail the same level of computational complexity that lateral analyses do, but unique features of each building's location can complicate the analyses. The following conversation at Seismic Specialists between Rush and Tim conveys how these complications arose. The engineers had been struggling for two days to come up with a pier solution for a building on top of rock. Their analysis kept yielding results that called for piers that were either too wide in diameter, too long in length, or too many in number to be practical.

Rush: [*dropping by Tim's cube*] What's the latest word? Have you heard from them? [He wants to know if the geotechnical engineers have called Tim.]

Tim: No. [*We walk to Rush's cube, where Rush sits in front of his computer monitor and types on his keyboard.*] Were you getting 50 feet yesterday? [*He is asking whether Rush's analysis stipulated piers 50 feet in length, which would be extremely long.*]

Rush: No, 100, it's ridiculous. [*He opens an Excel spreadsheet on his computer.*] These are their allowables [*he points to the screen, which shows requirements in a spreadsheet*]: 36-inch shaft, fill diameters are the same, flared to 5 feet, 30-foot pier is only 310 kips.[8] We're way out of the game here. [*He changes numbers in the spreadsheet.*] I'm missing a pi here, aren't I, here in my formula? [*He refers to a formula he has entered to calculate the rock capacity.*] This should be times pi, huh?

Tim: Yeah.

Rush: [*looking at the new values in his spreadsheet*] Now we're back in the game. Better check that. [*He looks at the terms in the formula bar.*] This is diameter times friction times end bearing pressure times pi.

Tim: What's that addition there? [*He points to a plus sign in the formula.*]

Rush: That's for the diameter. I just input it manually. [*He looks again at his result.*] It's better, but it's still not right. Don't trust what you do at the end of the day! Check it against their chart. [*They look at a printout of values that the geotechnical engineers sent.*] It is still huge. Try 25,000 for end bearing pressure. Triple the friction.

Tim: And this is straight into rock?

Rush: That's right.

Tim: So these are the shortest piers.

Rush: Shaft diameter is three and a half feet.... You're getting pretty heroic, a three-foot hole in the rock. [*He gestures with his hands, indicating that he means three feet wide. He changes the capacity formula again on his sheet while they discuss, and rule out, some other options.*] ... If you goosed the pressures, maybe you could use two 20-footers per column. That's all that we can know until we talk to these guys [the geotechnical engineers]. Will you check these so we know that they're right? [*He wants Tim to recheck the formula for rock capacity and the results it generates.*]

Tim: Sure.

The example of Rush and Tim shows how structural engineers must use their judgment when performing analyses. Here, Rush and Tim had to judge what end bearing pressure to use in their calculations. Uncertain what the correct pressure was, they concluded they might "goose" the value, meaning increase it, based on their best guess, to arrive at a reasonable solution. In the face of many unknowns, and with a numerical analysis that baffled them, the engineers had to make their design decision based on such assumptions.

In hardware engineering, limited tests of components at the block level are possible, but most often engineers need to test their components in the context of the entire microprocessor. Thus, integration in hardware engineering begins with each engineer independently creating an instantiation of the microprocessor on which to run his or her test. Each such instantiation of the microprocessor is called a "build." Creating a build means calling other engineers' code for the microprocessor's other components. In the firms we studied, this code lived in version control systems that strictly controlled who could make edits to the code for each component and which version of each component's code was available for others to use. Engineers routinely called for tens of thousands, and sometimes more than 100,000, code files from the version control system for their build.

Typically, the engineers created multiple tests at one time, and, depending on the firm, they also tested them on more than one build, or on more than one configuration ("config") of the microprocessor, at a time. To launch this array of tests, the engineers employed internally-created programs that organized and ran the tests and the configs in a testbench directory. It was not unusual for a set of tests to take more than twenty-four hours to complete. Testbench reports indicated which tests passed and which failed. An engineer's first task after running tests was to check the results to determine which fails were serious enough to debug and which fails would probably disappear when someone else updated existing code files.

Debugging code that failed a test was a riddle that could take days to solve. Engineers would first attempt to solve these riddles on their own. They would use trace programs, history commands, change finders, logs, assembly code, disassembly code, programs that contrasted versions of the same file, and other resources to help them in their efforts. Frequently, the hardware engineers needed the help of a colleague to resolve problems when testing. Sometimes the colleague whose help they sought was the author of code that their code called (and where they suspected the problem originated). Often, the colleague whose help they sought was the author of their component's code in the previous version of the microprocessor, and who they hoped would remember some fine points about the component's functioning that they themselves had overlooked. Many of our observations featured these collaborative debugging sessions.

For example, one day at Configurable Solutions when Eric's test failed, he sent an email to Rick with the subject line "help" and a message that contained no text, just five directory paths to testbench results. Rick, who had written the code for Eric's component in the previous version of the microprocessor, followed the paths and began examining the results. After twenty minutes, Rick sent an email to Eric telling him he thought the problem was that Eric forgot to "integrate" (incorporate) some changes into his code. Fifteen minutes later, Eric showed up at Rick's cube.

Eric: So you think I need to do some integrates?
Rick: Yes.
Eric: Integrates are nasty.
Rick: You changed too much. Make sure all the shared tools are integrated.
Eric: Where?
Rick: P2 share. [P2 was the name of the microprocessor; P2 share was a directory.] I added query routines. [*He searches through code trying to find when he made the changes, thinking it was within the last 1,000 changes, and then, finding nothing, within the last 2,000 changes. Eric waits patiently for a minute or two.*]
Eric: You basically want me to integrate all of P2 share [all of the code in this directory]?
Rick: P2 share is part of it, but that's not where o2text is. I don't remember changing o2text for some reason, and maybe it just didn't work so I pawned it off on Randy.
Eric: Why is o2text an issue here? Do you use it to initialize things?
Rick: o2text ... doesn't that figure out where the create space dot dat file, and figures out.... [*His voice drifts off as he looks at code on the screen.*]
Eric: Yeah.
Rick: So it needs to understand at reset where the physical address space it needs to create to be the diag and the literals all that kind of stuff.[9]
Eric: Okay.
Rick: And so I think what o2text expects—
Eric: I'll just integrate it, okay.
Rick: Yeah, I did a quick integrate and it had bugs and....

Eric and Rick agreed to meet in Eric's office in fifteen minutes to do the integrate. They worked there to debug Eric's tests for the next hour, and were still working together on the problem when the observer left. Such long collaborative sessions were not atypical in our observations of hardware engineers. During them, one engineer typically sat in front of his monitor while the other stood beside him, watching the screen and making suggestions.

Sending It Off for Production, and Getting It Back Again

The work of design and analysis is not complete after engineers specify all the details. Rather, after engineers send off the finished design to the individuals who turn the product into physical reality, they often receive notices of problems. These problems speak to the gap between design (the drawings, models, analyses, and specifications) and reality (the built product). Quite often, notices of problems require inspection by an engineer and a new cycle of design and analysis.

In automotive engineering, engineers do not need to wait until the vehicle is mass produced to get feedback about problems in design. Engineers conduct a number of physical tests in the midst of the design process. One reason to conduct the physical tests is to validate computer simulation models. A second reason is to identify design problems long before production in the hope of preventing the production of thousands of copies of a part that, if faulty, would have to be scrapped, and preventing subsequent delays that would tie up assembly lines, postpone manufacture, and cause IAC to miss its market deadlines. Many of these tests, however, do not employ physical versions of the parts under design because creating single prototypes, or "one-offs," as they were called, is expensive. A more common strategy is to create one-offs only for the parts that are critical to testing the new design, and to use parts mass produced for a prior generation of the vehicle for the other parts.

This strategy is the one we saw Nick, a noise and vibration technician at the company's proving grounds, employ one day when conducting a test to evaluate engine noise on a full-size truck. The analysis engineer's simulation models had predicted that the engine noise would meet the lead vehicle architect's criteria; the engineer had requested a physical test to confirm these virtual results. On receiving the test request, Nick began assembling a vehicle for the test. He tracked down one-offs of the truck's chassis and sheet metal outer panels. He then found a one-off engine and paired the engine with a previous generation transmission. Nick made these choices because he knew that the body panels were likely to deflect engine noise but that the transmission was unlikely to affect the noise. Thus, it was fine for Nick to use a previous generation transmission, but he needed one-offs for the panels.

After running the test, Nick sent the results to the analysis engineer. The results showed higher decibel levels in the axle than the engineer had expected. The engineer was not sure whether the results indicated that his simulation model was bad or whether the combination of one-off parts and previous generation parts had altered the results. The analysis engineer faced an important decision: to run a new test with only one-offs or to wait until the design of all the parts was closer to being final, at which time more accurate physical prototypes would be available. This kind of choice routinely faced the engineers, who on the one hand did not like to wait until late in the design process to conduct physical tests but on the other hand knew that tests involving interim parts that were not exactly like the final parts would be only so useful. In this case, the analysis engineer had a hunch that the physical tests results were an aberration and that it was worth waiting. He hoped his gamble to wait would pay off.

In structural engineering, changes that come late in the game are as likely to be due to changes or problems at the construction site as they are to problems in the design. For example, one day Sally at Tilt-Up Designs sent her drawings for a semiconductor manufacturer's fabrication facility (fab) off for permit review by the county, but afterward the client wanted to move a door from its planned location in one wall to another, existing wall. Sally needed to determine whether the client could do so without adding concrete to the wall for strength:

Sally: [*to observer*] They're asking, "Well, if we move the door over here [*she points to a wall on a drawing*] can we get away with not adding concrete to the wall?" ... I'm going do a few checks to see if indeed that will work.

Sally leafs through a binder, mumbling, "I'm just trying to find some calculations." She takes out two pages from the binder and looks at them, and then writes the door dimensions on her calculation sheet. She looks at a 3′ × 4′ drawing, and then turns to a set of smaller, detailed drawings. On her sheet, she draws two rectangles to represent the two sides of the door, which she labels $\hat{5}$ and $\hat{6}$. She transfers some dimensions to her sheet, writes a formula, refers to older calculation sheets for some numbers, and then substitutes these numbers for variables in the formula. She uses her calculator and writes down the result. She pulls down a manual from the shelf, *Design of Reinforced Masonry Structures*, opens to a page titled "Rigidity of Shear Walls with Openings," and writes:

DETERMINE PANEL RIGIDITY W/ NEW OPNG.
ASSUME $\hat{5}$ & $\hat{6}$ ACT TOGETHER AS ONE PANEL

She flips a few pages forward in the manual and looks at drawings of the deflected shape of a shear wall. She looks at the binder of old calculations and writes a formula for the deflection of the solid wall:

$\Delta_{\text{solid wall}} = 4[H/L]^3 + 3.0[H/L]$, where $H/L = 33'/48.1' =$

She calculates and writes the first part of the answer: 0.686. She looks in the *Masonry* manual, which shows an example of this calculation. There she looks at a drawing of a wall with a door and window, as well as some other formulas. She continues in this manner, looking at documents, writing formulas, finding values, and calculating. Finally, at the top of her third calculation page, she writes:

CHECK TO SEE HOW MUCH CONCRETE MUST BE ADDED IF STIFFNESS TO REMAIN UNCHANGED

After another couple calculations, she concludes:

1¾″ WOULD BE REQUIRED TO MAKE PANEL STIFFNESS UNCHANGED. THIS IS UNREASONABLE AMT. SINCE 1¾″ DOES NOT ALLOW ENOUGH ROOM FOR DOWELS, ETC.
SINCE 1¾″ UNREASONABLE, BUT STIFFNESS ONLY CHANGED BY 2% W/OPNG (& NO ADDED CONCRETE), THEN OK TO PUT DOOR IN WALL @ LINE A.

In this example, Sally was lucky because the requested change did not alter her design significantly. The contractor could move the door to the other wall without adding more concrete. Other times structural engineers were not so lucky, often having to resolve newly discovered problems with creative solutions that took into account half-built structures.

In hardware engineering, chips become physical after a "tape-out," or when the final design is converted into a photomask and sent to a semiconductor fabricator for production in silicon. This process was rather distant from the hardware engineers we studied: Between their verified design and a tape-out came several other steps that involved other engineering groups. These other groups "synthesized" the design by creating a transistor file, and then optimally placed and routed all the transistors. Perhaps because of the extensive simulation and testing that the hardware engineers performed, and no doubt because tape-outs were rare events, we almost never saw the engineers dealing with problems that arose after the product had been taped out.

An exception occurred one day when Eric at Configurable Solutions was approached by two software engineers. The software engineers had a printed circuit board with them.[10] They were using the board to test a new function (called "byte enable") that they wanted, but the function was not working. Eric had done the logic for the component in question (the system bus), but he was unwilling to provide the new functionality in the absence of a formal specification. Nonetheless, he agreed to help the two software engineers in the lab for more than an hour as the three of them used a logic analyzer to test the board and figure out why byte enable was not working. Eric knew that the software engineers' testing was insufficient to reveal that the quick fixes they wanted him to provide were not comprehensive enough or robust enough to appear in the microprocessor. In other words, Eric knew that more problems would occur in the future, even though the software engineers were telling him that everything would be fine if he would just make the changes they were requesting. Ultimately, Eric stuck to his demand for a specification.

Universal Themes in Product Design Work

Although the specifics differed, the general outlines of work were the same across the three occupations we studied. In each case, work progressed from rough ideas of what the product should do and how it should look to detailed specifications (derived from repeated testing) for each component of the product. The engineers associated with the beginning of the process—AVDC engineers in automotive engineering, principals in structural engineering, and chief architects in hardware engineering—enjoyed a certain status as compared with the much larger ranks of engineers who came up with component specifications. The division of labor did not stop there. Teams of engineers divided up the design and analysis work, and then later recombined their individual efforts into an integrated whole for further testing and analysis. In the course of doing this work, engineers within a team employed the same technologies as their peers doing similar tasks, but they used these technologies on different product components. Although engineers had individual assignments, they frequently collaborated, particularly whenever one of them had a problem in his design. The constant reworking of the design in the face of failed tests and new analysis meant that, although the design progression was linear and forward, the actual design process featured many loops back to earlier stages and technologies.

Also similar across occupations was the availability of advanced information and communication technologies for engineering work. In each occupation, engineers could choose from a range of sophisticated computational, logic, and graphical computer applications to aid them in building, simulating, analyzing, and visualizing their product designs. Each technology provided digital output, allowing engineers to store, transfer, share, reuse, and alter files at will. Every engineer worked at a desk with a computer through which he or she could access the technologies of his or her occupation, and each firm in our study kept up with recent advances in technology, evaluating and purchasing new technologies as necessary and desired.

A third similarity across the three occupations was the complexity of the products. Each product consisted of tens of thousands, if not hundreds of thousands, of components. The engineers we studied designed some of these components, while others came from external suppliers; all had to be tested as part of the completed whole product. As our descriptions of engineering work in this chapter illustrate, designing and analyzing complex products consumed the working hours of engineers in our sample. Theirs was not an easy task.

Despite these many similarities in the type of work, the type of technology, and the complexity of the product, these three engineering occupations differed greatly in their technology choices. In the next three chapters we describe in detail these differences and explore the occupational factors that shaped engineers' technology choices.

3 What Product Designers Let Technology Do

Which tasks belong to the human and which to the machine? According to the logic of a Fitts list, as we discussed in chapter 1, the way to determine what tasks a machine should do and what tasks a human should do is first to make a list of the strengths of machines versus the strengths of humans and then use that list when considering the requirements of the task at hand. For example, if the task calls for precision and repetition, machines are probably a better option than humans. If instead the task demands intuition or moral judgment, humans are the wiser choice. In practice, however, choices about how to allocate work between technology and humans are rarely as straightforward as a Fitts list might suggest. The allocation may be complicated, for example, because the task requirements are incompletely specified, or because the choice has ramifications beyond a single task. More broadly, occupational factors are likely to shape allocation choices.

In this chapter, we explore engineers' choices in allocating work to technologies—what we call the human-technology relationship—in the three occupations we studied. As we know from chapter 2, the type of work, the type of technology, and the complexity of the product are highly similar across these three occupations. The technology choices that the engineering occupations made, however, differed significantly. Choices within each occupation (across firms or groups), on the other hand, were identical. We show how occupational factors shaped the engineers' choices of when, why, and how they allowed technologies to help them. Focusing on the computer and its many applications, we find that different sets of occupational factors molded different motivations for engineers and shaped different choices of what to allocate to the computer in each occupation. Logics of understanding, speed, and belief motivated structural, hardware, and automotive engineers, respectively. These logics led engineers to minimize, maximize, or balance the role of the computer across the three occupations.

Structural Engineers and the Logic of Understanding

The structural engineers we observed were motivated by a logic of understanding when making technology choices regarding which, if any, tasks to allocate to computers. As structural models of buildings grew in complexity and detail over the course of a project, the structural engineers maintained vigilance in understanding, for each new step, what assumptions they were making and how those assumptions affected their models. The wrong assumption could lead to a feasible computational model with acceptable performance results in theory but a failed physical structure in reality. That computers in particular could yield "good" mathematical results yet lead to poor structures gave engineers, especially senior ones, considerable pause when contemplating which tasks to relinquish to computer applications. The logic of understanding thus prompted structural engineers to minimize the role of the computer in their work.

In fact, senior structural engineers were unified in their belief that the prevalence of computer analysis in university engineering programs had diminished the ability of recent graduates to think deeply and clearly about their models. Sam, a senior engineer in his early fifties and one of Tall Steel's four founders, was typical in his opinion when he lamented,

> Real engineering is done without calculators or computers. Computers are good for complex problems; they can confirm the answers that you anticipated, and thus can be used for testing solutions to complex problems. Virtually all engineers at all levels coming out of school today focused on computers and lack the ability to think.[1]

Sam provided an example of how young engineers failed to think through computer solutions in a case in which the county permit office had asked him to peer review a design submitted by another firm:

> I could tell just by looking at it that it was a poor design. They [the county permit office] sent me reams of computer printouts from it, which was ridiculous, a great big stack. [*He extends his hand about four feet off the ground.*] But all I needed to do was look at the drawing. [*He sketches in red pencil on a sheet of paper to demonstrate what he saw.*] For example, there was a place where a beam came in at an angle, like this [about 45 degrees], and it was huge, because the engineer said it was taking all the load because it had a fixed base. But beside it was this other member, and it was really skinny. And the thing is, the software kept telling the engineer to make this member bigger, and it kept taking more of the load, so it got bigger and bigger and this other one got really skinny. But that was all based on the assumption of the fixed base, and in reality it wasn't fixed at all, in fact this thing was sitting on almost nothing, so all the load really had to go through that very skinny member, which

couldn't take it.... It all happened because some young kid out of college learned the software, and just did what the software told him to do.[2]

Likewise, Roger, the president of Seismic Specialists and a man in his sixties, remarked,

There is a whole layer of analysis and computer activity that dominates time in school to the detriment of real-world, hands-on lab work in wood, steel, and concrete. When I was a student, we designed, built, and tested.... Now people in every school concentrate on analysis.... Now it takes longer to get people up to speed. The training is tedious, one-on-one with twenty-eight engineers. Analysis should be 10 percent of the job, which means the typical student is only prepared for 10 percent of the job upon graduation.

Yet despite senior engineers' reluctance to rely on computers, the increasing availability of desktop computing solutions for analysis had transformed the very set of tasks that structural engineers performed, broadening their domain from simply design (which elements of what material to choose and how to arrange them) to design paired with analysis (what size and weight the elements should have). Anwar at Seismic Solutions explained how this change had come to pass:

Up until fifteen to twelve years ago, companies the size of my firm, the ones that are anywhere from a one-man firm up to twenty-, thirty-, maybe even sixty-man firm, they did not do analysis. Structural engineering has two parts. One is to analyze a structure, to find out the forces, to find out the different actions that the structure needs to be designed for. The second part is actually designing those parts.

The engineering companies used to have a lot of designers and designers interacted with architects. They got information about geometry and they started drawing things. The most commonly used tool in the office was rulers and pencils.

At the same time, they needed analysis; they needed to quantify the size of the beams, what kind of forces they had to design the beams for. We used to give this task to outside companies. Often times the first round of analysis was very approximate and was done by hand calculators inside the office: they would get a rough idea about the size of members, and then they would contract a company that was specializing in analysis and they had access to mainframe computers. They did the analysis, the actual analysis.

The programs that were used in those times were notoriously user-unfriendly, and the system worked because the person using that was specialized in that program. Oftentimes that company wrote the software, in Fortran code or similar codes, C or Pascal.... They would get the drawings from an engineering company; they would run the analysis; and they would give that to the company to size the different members.

Since the advent of personal computers and since the new graduates are all coming out of the university with knowledge of how to use computers, analysis has been shifted from outside companies to in-house. You are doing that work inside.

Senior engineers' consternation in the face of this encroaching computerization led them to demand that their junior colleagues employ only technologies that provided transparency as to what the computer models were doing. Senior engineers held that designers needed to "see" what the applications were doing so that the designers could understand how and why their models were behaving as they were. Because many commercial software applications provided little transparency to users, engineers ended up having to do much of the work "by hand." That is to say, engineers did much of their work with little more than a pocket calculator, a pencil, and a tablet of green engineering grid paper. In addition, engineers turned to design manuals, textbooks, or calculation sheets from a prior project; these resources provided templates for how to set up the model and solve its associated equations. Such templates were typically "hidden" in any computer application, constituting the set of algorithms that the program enacted when asked to resolve a model. Thus, only a "by hand" method made visible to the engineer the calculations of the model as well as its assumptions, form, and content.

We saw in chapter 2 a clear example of how structural engineers worked "by hand" when Sally determined whether or not her client could move a door from one wall to another to avoid adding concrete reinforcement. Sally did not turn to a computer to aid her in her determination. Rather, she first copied information about the size of the door from her earlier calculations. Next, she opened a manual, *Design of Reinforced Masonry Structures,* to the section on the rigidity of shear walls with openings. From that manual, she learned that she could assume that the two panels of the door would move as one. She wrote down and solved with her calculator a simple equation, an example of which, with a related drawing, she found in the manual: $\Delta_{\text{solid wall}} = 4[H/L]^3 + 3.0[H/L]$. She continued in this manner, solving a handful of equations, until she had reached her conclusion: The door could be moved to the other wall without adding concrete to it because the stiffness of the wall would decrease only slightly with the addition of the door. In carrying out this task, Sally used only a pencil, an engineering grid tablet, her calculator, her old calculations, and the design manual. She did not use her computer at all.

Sam, the founder at Tall Steel, explained to us why structural engineers were better off forsaking computer programs, relying instead on approximations and straightforward formulas that they could easily compute in their head or with a calculator, much in the manner of what Sally had done. Sam's comments came as we were going over a list of technologies with him that we had seen at Seismic Specialists but not in his firm:

Researcher: What about Biax [a software application], which is for the nonlinear analysis of single members, as opposed to a whole system?
Sam: We never use that.
Researcher: You don't have it?
Sam: No, we don't have it. I mean, we rarely look at biaxial loading.
Researcher: What do you use instead?
Sam: [*chortling*] M over 4D! [*He means they use a formula, M/4D.*]
Researcher: So you use equations and work it out by hand if you need to?
Sam: Yes.
Researcher: What about Drain [another software application], for nonlinear seismic analysis?
Sam: I used it before. We had an employee who was a PhD and a specialist in Drain. It is very hard to use. We used it for push-over analyses of bridges.
Researcher: Meaning, can you push a bridge over?
Sam: Yes, for the push-over forces on a bridge. But it is limited to the types of analysis it can handle, and designs have to be "bastardized" to fit the program. So that is the main reason we don't use it, not just because I don't know how to use it and so then we don't use it.
Researcher: What do you use instead?
Sam: We use pencil instead.
Researcher: Can you tell me what is meant by "approximate analytical method"?
Sam: [*smiling*] M over 4D is an approximate analytic method. [*He takes a long pause.*] When I first got out of school they were doing high-rises all over the country, so I spent a lot of my time doing conceptual design on high-rise buildings, and you get something like that [*he sketches a tall rectangle with several horizontal lines denoting stories of a building*] with lots of members, hundreds and hundreds of members. Okay, very complicated, and you're going to stick it into SAP2000 [a software application] to analyze, right? With a load on top [*he draws a force arrow at the top of the rectangle*].

Well, it turns out that that's pretty much the same as a beam with a load on the top [*he draws a beam to the left of the building, whose height matches that of the building*] and I know that the deflection of this is PL^3 over 3EI. If this ratio of height to width is more than 5, then the behavior of this frame is dominated by the column behavior, which is approximated by this [*he points to the simple beam*]. So I can do a calculation that in ten seconds gives me the deflection of a sixty-story building within 10 percent, given the size of that column.... That's an approximate analytic method....

And after we're all done, then we stick it in the computer, and the computer does a very complicated analysis and comes up with the same answer. And approximate methods are very fast; they're very transparent; they tell you exactly what ... where your sensitivities are, and that's why we use them for designing the building to begin with. SAP2000 does a very precise [*he draws these two words out*] numerical analysis of the structure. It solves all the simultaneous equations. It gives you a precise answer. But it's very difficult to iterate, and it's nontransparent. You can't tell what contributed to ... all it tells ... it gives you one answer at the top of the [building] ... you can't tell what anything had to do with that [answer] at the top.

The approximate analytic methods of which Sam spoke were indicative of the field's ability to reduce the complexity of the product through simplifying assumptions. These simplifications made design and analysis possible, and in particular made it possible for engineers to eschew computer programs. By using their heads, in a manner of speaking, the engineers could use their hands.

Structural engineers also reduced product complexity by working almost entirely in only two dimensions. We were particularly struck in the field by the structural engineers' lack of enthusiasm for the 3-D capabilities of CAD. We had expected we would see them routinely using CAD to visualize their buildings, rotating them in virtual space to gain better perspectives of their design solutions than paper drawings could afford. At the time of our study, academic researchers were developing new technologies that would add yet a fourth dimension, time, to CAD representations. Adding time would render 3-D spatial representations as simulations that would display the construction of the building one layer or one system at a time. For example, simulations would show the pouring of concrete foundations, the construction of the walls, the addition of plumbing and wiring systems, and the like. But the structural engineers in our study expressed only a passing interest in such technologies; they found little use for a third dimension, let alone a fourth. Only on one or two occasions did we see a drawing in a 3-D perspective, and never did we see any engineer working in 3-D in CAD or any other technology, even when the technology had that capability (as in the case, for example, of Risa-3D, an analysis application). Several engineers told us that 3-D was merely a "fancy marketing tool" intended for architects when working with clients, not for real engineering.

The reason why structural engineers rarely needed a third dimension was that most of their elements were linear: beams, columns, struts, braces, frames, and other structural components followed a straight line, never zigzagging right or left or diverging up or down. Linearity of elements meant that structural engineers could easily visualize the third dimension by simply extending what they saw in the two dimensions of their paper drawings, almost all of which featured plan (from above) or elevation (from the front) views. Although avant-garde buildings, such as Frank Gehry creations, require 3-D visualizations to handle their unusual intersections of design components, the buildings that the engineers in our study constructed were mostly everyday rectilinear designs, for which 2-D visualizations were more than sufficient. Reducing product complexity by eliminating a dimension freed these engineers from a reliance on computer technologies.

Having confirmed their understanding of the model through an analytical method like the ones Sam described, or having completed at least one round of the full calculations on their own by hand, the structural engineers next operated with a calculus of efficiency. According to this calculus, if an engineer had a large number of similar calculations to complete, she would boot up a specialized software application or a general-purpose spreadsheet program. If instead she had just a few calculations to complete, she would take out a new sheet of grid paper and use her handheld calculator. Sheila, an engineer at Seismic Specialists, explained:

Researcher: What decides for you when you will do some analysis by hand versus when will you use a program?

Sheila: Usually it depends what type of project I'm working on. For some of the smaller wood buildings that I've worked on, we don't have very much software for wood design at all—I don't know if there's any commercial software. But usually ... I'll do it by hand to make sure I understand it and then I'll put it into a spreadsheet because there's a lot of repetition, and I don't want to re-do it by hand every single time. If there's something I'm only going to do once, I'll do it by hand.

Structural engineers strayed from this rule of first doing computations by hand before assigning tasks to the computer only when the computations facing them were too complex to do even a few by hand in a reasonable amount of time. For example, lateral analyses of buildings, which examined not the effects of the downward forces of gravity but the more complicated effects of the sideways forces of wind, earthquakes, and the like, involved FEA computations with differential equations. Completing such computations by hand was theoretically possible but in general not feasible, because doing so could take six months or more for a typical building. Hence, engineers allocated lateral analyses to specially designed software applications. They often preferred to begin by running a problem with a known solution through the application to confirm that the application generated a correct answer. They also scouted out documentation on the application, such as an instruction manual, so that they could review step by step the algorithms that it employed.

That was the case one day with Tim, an engineer at Seismic Specialists. We introduced Tim in chapter 2 when he and Rush were designing piers to place into rock to support a building. On another day on that project, we saw Tim trying to employ a computer application called PTData, a program for the design and analysis of post-tension beams. Tim ran into problems because he was unable to click on (select) within PTData the feature that he wanted to employ for the analysis of prestress losses. This feature appeared on the screen as an option button with lightly shaded gray text; the other options, which he could select, had black text. Tim was unsure whether the option he wanted was shaded gray because his firm had not paid for it or whether the application automatically employed the underlying algorithms, making its selection unnecessary. He doubted the latter was true because the forces should have been higher at the ends of the beam than at its center; when he looked at the number the program generated for the force at the center of the beam, it was too high.

Tim read the relevant chapter of the software's manual but did not find his answer. He was paging through the manual when Curt, another engineer on the project, stopped by.

Curt: Have you done runs on PTData? Have you done the prestress losses?
Tim: Do you know how to get into it? [*He means the shaded button area of PTData that will do the analysis of prestress losses.*] I can't get in. [*Tim and Curt look at Tim's computer screen.*]
Curt: I think it does it automatically.
Tim: I can't get into it. The effective force.... I guess I'm having trouble understanding....
Curt: Have you talked to Philip about it?
Tim: No, Philip was busy. But, you know, through friction and wobble the tendon ... [*his voice trails off*].
Curt: I would assume that they take that into consideration. Umm ... it might be in the manual.
Tim: Yeah, but I looked in the manual and it doesn't mention anything about it being an extra feature of the program. Actually do you know ... [*his voice trails off*].
Curt: Well, effective free stress—doesn't that take into consideration the losses?
Tim: It's the actual ... no, this is just beyond number of strands times the area of the tendon times the maximum allowable for the tendon. [*He grabs his handheld calculator and begins punching in numbers that he reads from the hard copy printouts. They both look down at what he is doing.*] This force here ... divided by the number of strands ... [*inaudible*] the area of the strands. You get to the 173 stress. So there's no reduction in the center span, so the prestressing works.
Curt: You're probably going to have to talk to Carl [a principal in the firm] on that issue because I don't know exactly what the program does.

Tim: Okay.
Curt: So you're worried that you got less prestress in the center spans that this isn't accounting for? [*They return their gaze to the screen.*]
Tim: Yeah. I was just wondering how to get into that portion of the program and see what your reductions really were to the prestress portions of the center spans.
Curt: Well, I know it does take any one particular span and divide it into ten sections and looks at each of those sections, but whether it looks at the effect of prestressed.... Do you have an output for a run that we can look at? I mean, is this all the output? [*Curt wants an output so that they can work backward from a solution to check what the program does.*]
Tim: This is all the output I've gotten. I'm just starting to learn the program so.... It seems to format it kind of funny, too.
Curt: It's terrible the way it formats, it doesn't even tell you what run you've got, you know. It looks like they took a DOS version, Window-fied it.
Tim: Put a Windows interface on it.
Curt: They didn't do a very good job with it.
Tim: It's odd how you have to go to this separate program to print it all out.
Curt: Yeah.
Tim: You can't print it from the main program. This is I think....
Curt: Do you have the manual?
Tim: Yeah. I printed out this. [*He grabs the manual from his desk.*] Philip said there are a bunch of sketches....
Curt: Yeah, we haven't been able to get them because they [the software vendor] got them in some different program, so we've called them, Philip called, and told them to fax us a hard copy because we were wasting too much time trying to get it from them.... [*They look at the manual.*] Is this the input part?
Tim: Yeah. I just printed out all the Microsoft Word files they had in the installation directory.
Curt: Can you pull up the ... [*They examine the large bar menu options, all of which have black text except one, which is shadowed out, indicating that they cannot access it. That bar is labeled Variable Pre-Stress Option.*] Is this the output?
Tim: This is run; it's being calculated.
Curt: OK. So you can pick any of these as we.... Look under, um ... I don't know which one. It's one of the devices under the ten sections.
Tim: [*pointing to various options on the screen*] Right here for the moments, span ... span moments, these are unfactored.... This is divided into ... [*his voice trails off*].
Curt: OK. That's the one I was thinking of. That doesn't tell you.... The only thing I can think of is to go into the theory part of this manual.
Tim: I don't think I have the theory part here.
Curt: OK, I've got it in my desk. You want to come borrow it?
Tim: Yeah.

They go to Bill's desk, which is back behind where the drafters sit. They get the theory manual, a photocopied document 8.5 inches by 11 inches with a cover.

Curt: Any other questions?
Tim: I'm doing the beams. Seems to work. I'm still trying to understand the algorithm [in PTData]. See what it is doing.
Curt: Talk to Philip.

Just then they run into Philip. Tim begins to question Philip in the aisle, and we make our way back to Tim's cube, where the three men stand inside. After a bit of discussion in which it becomes clear that none of them understands how the program treats prestress losses, Philip offers Tim a previous solution from another project so that Tim can go through it step by step to try to glean understanding.

Tim's example, like the others we have presented, illustrates the logic of understanding that motivated structural engineers. We discussed how the low transparency of structural engineers' technology helped shape this choice: Because engineers could not readily determine how software programs operated, they were hesitant to assign tasks to the programs. Adding to structural engineers' hesitation was their inability to fully test their designs.

Unlike automotive engineers, structural engineers had no laboratory or proving grounds where they could submit their buildings to physical experiments and tests. No firm in our study ever built a prototype building and then shook the ground around it to see if it would stay up or not. No firm had a room in which engineers tried out combinations of beams, columns, and connections, testing failure under different loading conditions. Rather, as they told us on more than one occasion, nature was their laboratory. After every earthquake, structural engineers eagerly awaited reports of building damage, which they analyzed to determine the source of failure. Because engineers had no ability to test their designs physically, they placed a high premium on understanding their models' assumptions and outcomes. With little faith that computer applications would take the necessary care and precaution that they themselves evinced in establishing these assumptions and examining these outcomes, they chose to do the work themselves.

Liability concerns and government regulations also played a role in shaping the choice to minimize the computer's role. Anwar, a Seismic Specialists engineer, explained, in the context of designing beams, how these concerns kept structural engineers from allocating tasks to computer programs:

I would say the design of beams cannot be made any simpler. It is fairly low-tech, mindless work. The only thing is that there is a lot of it. There are a lot of beams. They all need to be checked. So what we used to do by spreadsheet in the past, Ram-

Steel [a commercial software application] does right now automatically. You just define the loading criteria and the program designs that for a thousand beams at the same time. One of the issues with engineering, structural engineering especially, is the idea of liability: we still have to go back and check the results, make sure that inadvertently a mistake has not been made. Because of that there is an inherent reluctance to use a canned program.

Structural engineers' reluctance to use computer programs was universal across the three firms that we studied and evident in how the engineers approached the task of checking shop drawings. Shop drawings are the production drawings that steel fabricators create based on the engineering firm's drawings to guide the manufacture of all the beams and columns to be used in the building. The engineers deemed the task of checking the steel fabricators' work boring, but they did not allocate it to a computer program because they were liable for making sure that the design was correct. Most mistakes in the shop drawings were the fault of the steel fabricator, not the engineering firm, but engineers checked the drawings carefully nonetheless because if mistakes occurred, sorting out who was at fault and who had to shoulder the cost took time.

Ivan, an engineer at Tilt-Up Designs, demonstrated this concern one day as he set out to check a set of shop drawings:

Ivan: [*to the observer*] Let me explain. I don't think you will be very interested in what I am going to do today. I will just check shop drawings. You know, I am PhD. I have PhD from [an Eastern European country]. I worked there on many projects. But here, I do not have my license. So today, I will check shop drawings. You are welcome to watch, but I don't think you will find it interesting, because there is no technology. Just the drawings, a pen, that's it.

Ivan rolls out some shop drawings on his desk. With a red ballpoint pen, he circles items by making clouds with his pen. He says he finds the work very boring. He writes the word "verify" in many places.

Ivan: I think the manufacturer has specified the wrong size [of a beam]. So I tell him to verify.
Researcher: And how will he do that? What will he check them against?
Ivan: The original drawings. [*Ivan points to a bound set of engineering drawings on a nearby desk.*]
Researcher: So it is not your responsibility to do so?
Ivan: No, it is his responsibility. Some of the mistakes are not so bad. [*He points to an upside-down label on a beam.*] This is not a problem. It is wrong, but no one will make a mistake from it. But here [*he points to another area*], I must show this, the beams must be lower here. How this mistake occurred? [*He refers to some beams that*

are mislabeled.] Ok, they copied the label in CAD, and forgot to change it. It should be C8, but it reads C9.

Researcher: Ivan, how much of the task is just copying versus using your knowledge each time? Could someone else copy over the information for you?

Ivan: I am not just repeating. Each time I must think. [*He begins a new set.*] This time I start from the back and work forward, the opposite order. If I do it the same way, I might miss something. So this way I have to think.

Ivan's concern for detail and accuracy in the task of checking shop drawings stands as testament to how strongly liability concerns and government regulations shape technology choices in this field. As Sam from Tall Steel told us, allocating the task of checking shop drawings to computers was technologically feasible, but not viable: "The challenge in doing so is not a technological one at this point, but a political one." If concerns about liability and government regulations kept structural engineers from turning over even the most tedious and repetitious work to a computer program, then all the greater their worries must have been for work that required greater judgment, knowledge, and intuition than did the task of checking shop drawings.

Several other factors facilitated structural engineers' choice to minimize the computer's role, though they did not strongly shape that choice. Structural engineers could afford to do many calculations by hand because once they had finished bidding on a project, they did not need to rush to bring a product to market ahead of the competition. Instead, they worked to deadlines. Although engineers often felt pressured to meet their deadlines, in general they operated without a compulsion to work quickly. In this sense, the fact that competition in this occupation is focused on bids and not on market share permitted, but did not force, the engineers' choice to minimize the computer's role. Even without a need for haste, the engineers could have decided to allocate more tasks to the computer had they so desired.

Similarly, the structural engineers' choice to minimize the computer's role was facilitated by the fact that each design they did was unique. Theirs was a custom market. The uniqueness of designs that were tailored to the needs and desires of each client in this custom market meant there was little impetus to digitize solutions for reuse, and hence little need to create and store digital artifacts. Granted, each firm kept databases of standard connections and other partial solutions that were common in many buildings, a practice that kept them from reinventing the building, as it were, with each new project. But because no two clients ever wanted exactly the same building, no two designs were the same. Thus, the structural engineers did not begin a new project with a corpus of drawings from a past project that they intended to modify to create "the next version" or "the latest model" of the building in that way that hardware engineers and automotive engineers designed their new products.

The fact that one building's design relied little on the design of any buildings before it is suggestive of another occupational factor. We call that factor product interdependence, or the degree to which one product is dependent on another product. In the case of structural engineers, because each building's design was independent of, and different from, every prior building's design, product interdependence was low. This low product interdependence acted in combination with the custom market to aid the structural engineers' choice to allocate tasks to humans, not computers.

The structural engineers' choice to minimize the computer's role in their work had two direct consequences in the workplace. First, this choice meant that structural engineers operated in a physical world of work in which engineering grid paper, hand calculators, sketches, pencils, erasers, liquid Wite-Out bottles, staplers, hole punches, three-ring binders, and engineering drawings in a variety of sizes covered their desks and filled the recesses of their cubicles. We see evidence of this physical world in the detailed description of Sally's cubicle as well as in the examples of structural engineering work in this chapter. By and large, physical artifacts in the structural engineering workplace were the remnants of work done by humans, not computers.

Second, because the human was such an integral figure in the carrying out of design and analysis tasks in structural engineering, little could be accomplished without humans. Specifically, structural engineers could not launch computer programs that would run indefinitely in the background while they completed other work. Instead, because engineers assigned only small tasks to computers, computers finished their tasks quickly, leaving engineers little time to begin other work on or off the computer. As a result, structural engineers almost never multitasked. They were at times interrupted, having to switch from one task to another, but we never saw them simultaneously performing two tasks beyond the routine fare of office work, such as clicking options on a menu screen while picking up a ringing phone. With the computer playing a bit role, the human held center stage: equations got solved because humans solved them, drawings got changed because humans changed them, and models got made because humans made them.

Hardware Engineers and the Logic of Speed

Unlike the structural engineers, whose concern with understanding their models drove their choice to minimize the computer's role, the hardware engineers' main concern was how to do their work faster. Motivated by a logic of speed, and recognizing that computers could do the work more quickly than the engineers could, the hardware engineers opted to maximize, not minimize, the computer's role in their work. Therefore, the hardware engineers assigned as many tasks as they could to their computer technologies. Raymond, a hardware engineer from Programmable Devices, evinced this motivation to allocate tasks to computers when talking about the task of verifying netlists[3]:

> In netlists, we want to go through and see which nodes are utilized, which are not. Tools can't do that yet.... We're working towards more [use of technologies]. What can be done, will be done.... It's not that no tool is out there. There might be one, but it's just that we have no time to look. We're working with the CAD group, they let us know; they provide us with documentation on new tools, and we look at it and see if we want to pursue it.

The need for speed had ramifications extending to the smallest of tasks. For example, the engineers in one firm went so far as to assign the phone directory to a computer program. To look up someone's number, they simply typed, say, "Phone morty" at the Unix prompt to get the number of a fellow whose last name was Mortensen. At the opposite end of the scale, engineers allocated the largest tasks to computers at the end of the day, so that computers could complete the work overnight while the engineers slept at home, with the results awaiting them in the morning when they returned to the office. The most common such tasks were simulations. Brian at Configurable Solutions explained his temporal allocation of simulation work, which he organized across a group of hardware engineers to ensure efficient communal use of the servers:

> I typically begin my day by logging on to my [simulation] account. I look at the results from the night before, then manually post the results onto the web page every morning. I often log in at home in the evening to make sure things are running ok. About three every day, or 3:30, I use the "at" command in Unix to schedule the tasks for the evening. There is a shell script that sets up the [simulations]. It goes off about seven in the evening. So all the RTL workers [hardware engineers] know that the deadline is seven.

Because hardware engineers allocated so many of their tasks to technologies, they typically had multiple technologies working for them "in the background" while they completed other work in the foreground. That is to say, the engineers multitasked by sending certain tasks off to a bank of servers for completion, and then doing other work that required more interaction, judgment, and decision making. The arrangement worked because the tasks that the engineers sent off to the servers were large enough to take many minutes, often hours, to complete.

In one example, Christopher at AppCore spent thirty minutes one morning reading a hard copy of *EE Times* while a program ran in the background, its interim results scrolling by on his first monitor. When the program completed (the scrolling stopped), Christopher launched a "make" command via that monitor to build a testbench. He then turned to his second monitor, where he began to enter comments into a different piece of code. After some time, a notice appeared on his first monitor informing him that the computer had generated code for eighty-seven of eighty-seven modules. At that point, Christopher launched a "diag" (test) on that monitor, and then returned to his second monitor to continue entering comments. Christopher's actions illustrate how having two monitors—one for work carried out at a distance by a bank of servers that needed little supervision, one for work on his desktop with programs that required his attention and input for progress to be made—facilitated his allocation of tasks.

With banks of servers working for them in the background, the hardware engineers were free to engage in a variety of pursuits, not all of which were work-related. Engineers at one firm were fond of playing foosball in the company break room at all hours of the day, not just lunchtime. Other engineers made personal phone calls or read online local or national newspapers while their programs ran. Managers who walked by and observed these activities never once commented on them.

Managers kept mum because these breaks, though noticeable and at times amusing, were not the engineers' most common use of their "free" time. Rather, engineers frequently used this time to talk in person with other engineers in the firm about specific problems that they were having. This précis from our field notes at Configurable Solutions was not atypical:

Fazio had been sitting at his computer working when he turned to our observer and said, "I am rebuilding the config [model of the microprocessor] with 'make.' It will take a few minutes. Meantime, I have to ask someone something." Fazio then walked over to a cubicle on the other side of the building. He told the cubicle's occupant, the sys admin, that he had completed his favorable evaluation of a new code-checking application and now wanted to make the application available to all the engineers in the firm. After making his request in person, Fazio returned to his computer, where he set up some tests to run on his completed config. Fazio repeated this sequence of activities several times over the course of the next couple of hours, beginning a task in the background, and then walking to someone's cubicle to discuss some issue. That afternoon, the issues included what documentation customers required, how a certain software application operated, where a set of digital files was located, and what Fazio's password was for a particular internal application area.

Another common use of the hardware engineers' free time was to write scripts designed to allocate more tasks to technologies by automating the links between technologies. In the next chapter, we discuss automation of this type in detail. For now, this comment from Raymond at Programmable Devices illustrates how he developed an interface that would allow engineers in his firm to set up a large number of commands for the computer to execute sequentially. His interface would relieve the engineers of having to type in each new command after the one before it in the sequence completed, as he explained:

> Our flow [sequence of design technologies] is all these gmake scripts and things you need to type. So we're just putting it all together so we can push buttons instead of typing all these commands. This is a project we're undertaking to develop this nice little interface. It goes through and does these gmake things in the background.

Two tasks that hardware engineers could not easily allocate to computers were writing and debugging code. To write code, the engineers almost always started with the prior version of the microprocessor code for their component and made changes. We never saw engineers starting from a blank screen. Starting with previous code had two immediate advantages, both related to speed. First, the bulk of the existing code typically would remain in the new version, so starting with old code prevented engineers from duplicating effort. Second, on those occasions when the engineer needed to overhaul much of the code, he still found it easier to refer to the old code for examples of formatting syntax than to look up syntax specifications in a code manual. Because hardware engineers employed a wide variety of programming languages, they could not remember the nuances of syntax for each one. Thus, with prior code in hand, the engineer would begin the long, involved process of writing new code, a process that demanded that he think through each line of code carefully to ensure that the component would function in the manner intended and would interact appropriately with the chip's many other components.

We encountered an example of writing code in chapter 2, when we described how Doug at Programmable Devices needed to optimally route the paths of two new clocks in the multi-gigabit transceiver (MGT) and BRAM (memory cache) components of the new microprocessor. Where we left off, Doug had determined that his task would require that he figure out a good path to BUFG, a buffer used to relay clock signals. In the précis below, we see how he found that path:

Doug: [*examining a schematic of the design on his screen, and pointing as he talks to the observer*] Now, a good route to the BUFG mux is pretty easy to see. The BUFGs are inside here, so if we just jump in here and take a quick look, these are the BUFGs. And the lines that come in here that I'm interested in are DLFOUT. So these are lines that go across the chip in this thin row that you see here. And that's what we want to get to, for having a good path to the BUFG.

Doug: So what I'm trying to establish now is, what's the nature of this connectivity and what should I do with the output clock signals to make it so I can get to this point? I'm just looking to see how I can get here efficiently from the transceiver. I want to trace these signals back because these—I happen to know they are coming out of the transceiver, and these are really the lines I'm interested in. But not necessarily all of these lines connect up to where I want to go. All this means is that at least one bit out of this bus is connected. And I don't really know a perfectly efficient way to do it, so I'm going do it by brute force. There are only eight of them, so it shouldn't be too bad.

Doug proceeds with his "brute force" method of tracing each of the eight paths "by hand." Afterward, he consults internal documentation that defines the connectivity from the transceiver. He wraps up his work with a conclusion:

Doug: What I've just established is that any of these lines is capable of getting up to the routing to the BUFG, which is an impressive result. I was expecting that only some subset would be able to do that. So that's the first interesting result of the day.

Doug's example shows how, although the hardware engineers used technology to help them in the task of writing new code, they did not relinquish the task to technology. In Doug's case, he used a visualization program called Composer that created schematics from his code to help him see the paths, which he traced "by hand" on the screen. Part of the reason why the hardware engineers did not assign this task to the computer was that the act of writing code involved much more than simply typing lines of text. In this example, Doug first had to determine which paths were available for his clock before he could began to type out the lines of code that would implement his design. By no means could Doug complete this task in the absence of computer technologies, but neither could he simply assign the task to a program that would run in the background. He had to first determine where his clock needed to travel, and then sort out how it might best get there. That kind of reasoning was not within the purview of his computer technologies. At the current state of the art, technologies could handle routine routing problems, but they could not handle problems in which optimal performance was a goal.

Similarly, to debug code, the hardware engineers had to rely on their thought processes, often trying to deduce from computer-generated evidence in trace logs and test results where their code went wrong. Although test results might have indicated, for example, that the code assigned a wrong value to a variable, and trace logs may have pointed out where the code had made the wrong assignment, no program existed that could tell the engineer how to fix the problem in a manner that was consistent with the desired functionality of the component. For that, a human was needed.

When debugging proved too vexing for engineers to handle alone, they turned to colleagues for help. Frequently, they would beckon a colleague to their cubicle, where the two engineers would hunch over a monitor, puzzling out the problem together. This précis from AppCore illustrates collaborative debugging:

Ramesh comes to Luke's cubicle and tells him that he was trying to do Luke's analysis of time delays on clock paths, but was having problems. He asks Luke if they can go to Ramesh's cubicle and look at his computer because he can explain things better that way. In Ramesh's cubicle, Ramesh sits in the chair in front of his computer while Luke stands beside him. He begins explaining two clock paths, but Luke interrupts and tells him that he will never use both of the clocks simultaneously. Then Ramesh asks how Luke accounts for time delays that are appearing on one of the paths. Luke confirms that time delays are confusing in Verilog, but says they are clear in VHDL, another language for creating RTL models. He does not understand why the time delay is occurring in the Verilog code given that there is no violation. Ramesh is also surprised. The two men continue on for several minutes, trying to sort out what Luke needs and how Ramesh might achieve it for him, until finally Ramesh comes up with a solution:

Ramesh: What I am saying is that I can create a virtual clock and then make this [the two clocks] both be derived out of that.
Luke: Yes, that would be exactly what I need, because I need the entire clock insertion delay problems that might arise from that.
Ramesh: Okay, let me set this stuff.
Luke: All right.

Luke returns to his cubicle. Forty-five minutes later Ramesh sends him an email with timing path delays detailed in it. While Luke is still reading the email, Ramesh drops in and they discuss what might be causing the delays.

This example from Luke and Ramesh highlights how the engineers needed to think through the problem (here, how to handle time delays when timing two separate clock paths), the issues involved (e.g., do violations disable a clock?), and a possible solution (in this case, creating a virtual clock from which the other clocks would derive). As the example suggests, debugging code was not a task that engineers could simply hand off to computers. Given the complexity of the microprocessor, debugging consumed a considerable portion of engineers' time and significantly detracted from their goal of speed, especially when some problems, like this one, required not one human but two.

The goal of speed in hardware engineering is in large part a product of the competitive landscape for mass-produced chips: Every firm we observed was in a rush to gain market share by being first to market with a new product. The hardware engineers were attuned to these competitive market dynamics: In public firms in our sample, hardware engineers, many with stock options in their firm, tracked in real time the firm's current market price via small browser windows perpetually open in the corner of their computer screen. The fear that someone would beat them to the market propelled engineers to assign as many tasks as they could to their computer technologies. Engineers at times blamed managers for pushing them to work fast. But engineers themselves understood and embraced the logic of speed, as the comments from this group lunch at Configurable Solutions demonstrate:

Colette: Never underestimate the power of a schedule.... And I think engineers are great at that: "Give it to me for yesterday." I think pretty much everybody—whether one comes up with a more elegant solution than another, or a more documented solution than another—I think they all will come up with a solution quickly. The industry is moving so fast that you have to move fast.
Rick: I tend to think it's more that engineers want to do the elegant, documented, big-picture idea, and it's the manager's job to beat the living snot out of them to hit a schedule.
Colette: That's what I say, "Don't underestimate the power of the schedule!"
Rick: Yeah, but I'm not sure it's that because of the schedule we desire to do it this way. I think it's where—
Brian: [*interrupting*] But you're not here to do research, you're here because you want this company to go IPO and make a lot of money.
Rick: Right.
Brian: You chose Configurable Solutions over another opportunity, at that company that was *losing money* [*he emphasizes the words*], where you could have done something really fancy and elegant, but they don't make money.

Eric's words below, spoken one April morning to explain his task to our observer, reinforced the notion that engineers felt pushed to work quickly.

His words also showed what happens when engineers had to forsake the elegant solution for the expedient one:

I'm going to add a hack. A hack is a temporary Band-Aid to a problem, but in reality they end up staying forever. [*Using the grep command, Eric searches for "XXX."*] If I look through a file, there are a lot of things with "XXX's," which is a marker that this is something temporary, that it will be changed. [*He reads the result of his grep command.*] There are seven-five lines of text with "XXX." The problem with engineering courses, they are not as time-bound as things are here. I have this thing due by June, but I have a lot of things due by yesterday. It requires time management.

Although the dream of making money spurred the hardware engineers toward speed and drove them to allocate as many tasks as they could to their fast technologies, they were able to do so only because they could fully test for failure. Unlike their structural engineering peers, hardware engineers did not need to worry that the impact of their modeling assumptions on the performance of their design would elude them. They were free of this worry, and hence able to allocate tasks at will to computers, because they lived in a binary world of ones and zeros where truth was self-evident: either their code worked or it did not. If the code survived a battery of tests—some planned, some random—and delivered at the end the right answers, then the code was good, period. Hardware engineers created elaborate testing schemes to ensure that they had tested for all possible failures, a situation they called "coverage."

Moreover, whereas the structural engineers could only ponder the inner workings of the commercial software products that populated their workplaces, the hardware engineers crafted much of the computer code they employed, and what they did not craft they could readily read and reconfigure as they chose. Rick at Configurable Solutions provided a clear example of this ability one day when he downloaded from the site of a version control system a "wrapper," or script that made possible the running of another program. In this case, the wrapper that Rick downloaded was intended to make his firm's commercially purchased version control system (Perforce) act like a different system (Aegis), available as freeware. Rick was particularly keen to adopt the enforcement policies of Aegis in Perforce. What proved fascinating about Rick's download of the wrapper was that, although the wrapper was written in Perl, a language Rick knew, an accompanying file, "myinit," was written in some language Rick did not know. Rick's lack of knowledge, however, did not stop him from opening the file in his text editor and modifying it. He simply looked at other places in the file for syntax examples to help him accomplish what he wanted to do. Rick kept making mistakes (e.g., using an equals sign when none was needed and using quotation marks when none were needed), but eventually, through trial and error and a bit more digging in the file, he figured out the correct syntax. That is to say, he programmed in a language he did not know, and in that manner he appropriated new technology, bending it to meet his aims.

If structural engineers pondered computer applications, then, hardware engineers probed, prodded, and perfected them. As a result, when using computer technologies to aid them in their work, hardware engineers enjoyed a transparency that structural engineers did not. Assumptions, in short, were quickly borne out or refuted; there was little need to worry whether models would act in opaque ways. All told, hardware engineers had strong motivation and clear opportunity to allocate much of their work to computers.

The result of this allocation was twofold. First, as we have mentioned, hardware engineers could multitask because their computers were working for them on one or more tasks in the background while they worked in the foreground on some other task. As a result, their workplaces appeared more freewheeling than structural engineering workplaces, with hardware engineers playing games in the break room, casually reading newspapers at their desks, and surfing the Web. Structural engineers, who could do only one task at a time and whose hours were billed to projects, engaged in no such frivolity. The difference in environment was less one of temperament than of technology choice.

Second—and this outcome was implicit in our examples—because hardware engineers maximized the role of the computer, their work artifacts were primarily digital. Consequently, their desks and cubicles were largely free of physical work artifacts beyond their phones and computer equipment. We recall from the introduction our description of Eric's office space, which was devoid of standard work artifacts like staplers, liquid Wite-Out bottles, and hole punches, though littered with personal mementos like yo-yos and bendable action figures. Hardware engineers spent a large portion of each day not poring over paper documents but facing computer screens and typing on keyboards, living as they did in a virtual world of work.

Automotive Engineers and the Logic of Belief

Structural engineers created objects that were too big; hardware engineers created objects that were too small; automotive engineers created objects that were just right—for holding in one's hands and inspecting physically, that is. Although structural engineers might have benefited from creating physical prototypes of the buildings that they designed, creating such models was infeasible for many reasons, not the least of which were cost and time. Consequently, the final building on the construction site was the first physical version of the building to be created. "Taping out" a computer chip to create a physical instantiation was similarly a costly and time-consuming enterprise. Moreover, hardware engineers would have found little use for physical prototypes of their chips because computer models could provide insights at granular levels that engineers would have been unable to gain from physical models, on which hundreds of thousands of gates and transistors would coexist in a tight area of less than a single square inch. Unlike their structural and hardware engineering counterparts, automotive engineers designed products whose size rendered the products highly amenable to physical prototyping. As a result, automotive engineers were prone to believing that which they could hold, see, feel, and smell, and they favored using computer applications in a manner that preserved this sentient knowledge.[4]

Automotive engineers, then, operated according to a logic of belief—belief in the physical over the virtual, belief that the virtual must be identical to the physical—when allocating tasks to computer technologies. This logic was perhaps most striking among the design engineers, who in the division of engineering labor worked directly with automotive parts rather than the entire vehicle. Design engineers were so oriented toward the parts that managers assigned to them that they often talked about "owning" the parts. A design engineer named Sunil expressed this orientation concisely:

> I own front brakes for midsize cars. That's my job. I design front brakes; I work with vendors to get those brakes built; and I work with the analysis engineers to get those brakes tested. I'm responsible for them. I think about brakes all the time. My wife makes fun of me. But I own the front brakes.

Sunil's cubicle reflected his feeling of ownership of front brakes and the strong physical presence they had in his work. Pictures of front brakes decorated his cubicle walls. The screen saver on his computer monitor was a continuously playing movie that depicted the construction and dismantling of a front braking system. Rotors, calipers, pads, and other physical brake parts that Sunil had taken from IAC's proving grounds littered his otherwise empty desktop. "I like to have the brakes here," Sunil commented one afternoon, "because it is helpful to see them and touch them. You can forget that they're real if you just look at your [computer] models every day."

In general, vehicle parts, like Sunil's front brakes, were relatively small in size. Thus, design engineers found it easy to pick up parts, carry parts places, and store parts on desks and in drawers. David, a design engineer, noted how the portability of physical parts allowed them to play a role that their computer representations could not:

I work on rear brakes and I think about brakes a lot. You can see [*he points to his desk*] I've got brake parts all over the office. I do it because it's useful for me but I also do it because I can. I mean, I can't bring in a whole truck into my office.... And so you've got the brakes here and you think about them, you play with them and they become embodied in your hand; they are not some abstraction like they are in the screen.

I sound all philosophical here, but it is really practical. I mean you've got to look at the brakes in real life because on the screen you just miss things or you don't think about things. Like [*he picks up a brake assembly from his desk and rotates it toward the observer*] you can see how tightly the pad is pressed against the rotor. Well, in CAD [in the computer representation], there might be a few millimeters' distance there, and you say, "Okay, sure, whatever." But then you see this [the distance in the physical part] and you say, "There is no way that distance can be there in the actual brakes," which is never something you would have known by looking at the [computer] model. So you are learning by looking at these [physical parts]. They make a difference.

A glance around an analysis engineer's cubicle, in contrast to a design engineer's cubicle, rarely resulted in sightings of parts. Analysis engineers, most of whom studied the performance of subassemblies or entire vehicles, did not stash parts on their desktops or tuck parts in desk drawers. One could not infer, however, from the absence of parts in analysis engineers' cubicles that physical parts held little meaning for them. Rather, that same glance around the cubicle would often reveal that the analysis engineer was not at her desk: She was off examining physical parts, grouped together as subassemblies or vehicles, in the larger spaces that these groupings required. Although analysis engineers used computer applications to simulate the performance of subassemblies and vehicles, they also spent a good deal of time handling physical parts at IAC's proving grounds, located about forty miles from the engineering offices in the design center. As Rachel, an analysis engineer who specialized in fuel economy, explained:

I was surprised when I started working here how much time we spend out at the proving grounds. It's not just watching tests, because you can watch the test on video and you mostly get what happened. But it's about inspecting the vehicles before and after the test, and looking at how all those parts that you work with regularly actually fit together. It is just helpful, and it's hard to explain why.

Trips to the proving grounds put analysis engineers like Rachel into regular contact with parts. Rachel, like most engineers we interviewed and observed, held a fairly myopic view on what parts actually "mattered" in the vehicle:

> You know what's funny? Sometimes I come back from the proving grounds—and I went and saw a car that was used in a crash test because that's the only one available that had one of the air cowls I used in my model, and I look at the whole intake system—and then I get back to the office and someone asks me if I saw the car and they'll ask what it looked like after the crash. Most of the time I can't even remember because I'm not looking at the whole car, I'm looking at the parts that I'm concerned with. Those are the only ones that matter. So I don't even think about the whole vehicle; I think about the areas I'm working on, and I know them inside and out.

Although analysis engineers at times had difficulty articulating to us why spending time at the proving grounds was important for them, our research shed light on why the physicality of the vehicle was central to this work. Our observations with Tushar, an analysis engineer who specialized in frontal crash analyses, provide a good example. Like most analysis engineers, Tushar used computer applications to test, by means of mathematical calculations, what happened to a certain part of the vehicle under particular loading conditions. In Tushar's case, this mathematical, computerized testing meant that he analyzed how the front of the simulated vehicle fared when it crashed into a simulated offset deformable barrier.[5] While Tushar was busy carrying out virtual crash tests on his computer, test engineers at the proving grounds were equally busy crashing physical vehicles into physical offset deformable barriers.

Ultimately, managers hoped that Tushar's simulations could substitute for physical tests at the proving grounds, which were costly in comparison to virtual tests. Ignoring the cost of test engineers and testing facilities, each physical prototype (called a preproduction vehicle) that engineers crashed at the proving grounds alone cost about $650,000, a number to be multiplied by the eight or so preproduction vehicles that test engineers crashed in the course of several years of design and analysis in each vehicle program. With IAC vehicle programs numbering in the scores, physical testing costs multiplied quickly. Thus, managers faced tremendous pressures to predict vehicle performance using simulations and to refrain from crashing real vehicles into real walls.

Computer simulations held other advantages over physical tests, including the ability to run many tests in a short period of time and to rerun the same test over and over again with the same vehicle. In the physical world, a vehicle once crashed was useless for further crashing because the engineers could hardly return it to its prior, pristine state. In addition, test engineers working as a team needed a good bit of time to set up and run a physical test, whereas an analysis engineer working alone could set a few parameters and launch a test with little more than a few keystrokes. Thus, the total potential savings of computer simulation over physical testing encompassed cost, time, labor, and other resources. Important in the competitive market for automotive vehicles, these potential savings provided a significant incentive to allocate work to computers.

Despite this incentive, managers understood the limits of the computer's potential role in automotive engineering. At the extreme, IAC managers were not prepared to discontinue physical testing (and, in so doing, allocate all testing to the computer) because government agencies in the United States as well other countries in which IAC sold its vehicles conducted their own physical tests to ensure that vehicles met safety standards. IAC did not trust computer crash simulations enough to fully substitute them for physical tests when government agencies would conduct physical, not virtual, crash tests. Instead, IAC wanted to maintain a few physical tests of its own to mimic government testing. In this sense, government safety regulations constrained the computer's role in testing.

During our observation one afternoon, Tushar was frustrated because the technicians at the proving grounds had sent him results from their physical test of a week prior that did not match his simulation results. Tushar was upset because the test engineers had employed in their physical test a version of his test design that was older than the test design on which he was currently working. Four months ago, Tushar had sent his test design to the proving grounds for testing. Building and instrumenting the vehicle per Tushar's test design and then fitting the test into their busy schedule had taken the test engineers four months. In the intervening time, Tushar had revised his test design, but had assumed his changes would not interfere with the physical test. Now that the physical test results were not matching his simulation results, Tushar feared that the last four months of his work had been rendered useless.

On receiving the physical test results, Tushar spent three hours going back through his computer simulation trying to find why the results it provided were different from those the test engineers reported. In addition to looking at the numerical outputs, Tushar placed a 3-D animation of the computer simulation on one side of his computer monitor and a movie of the physical crash at the proving ground on the other. He played the two videos at the same time more than two dozen times trying to identify what was occurring differently between them. At 5:00 p.m., he left the office still frustrated.

When our observer returned the next morning at 9:00 a.m., Tushar was already at his desk playing the two videos side by side again. He remarked, "I didn't sleep much last night worrying about this. I think we're just going to have to go to the proving grounds today and take a look at the test vehicle." Later that morning the pair arrived at the proving grounds. The technician whom Tushar had arranged to meet led them to the test vehicle, located in the corner of a large warehouse that contained at least fifteen other crashed vehicles. The entire driver's side of the vehicle was crushed; the rest of the vehicle appeared to have suffered no damage. The following excerpts from our field notes detail Tushar's method of exploration:

We approach the vehicle and Tushar puts his hand on the hood. He gently sweeps his hand across several creases in the steel and then steps back a few feet to view the vehicle in its totality. He squats down and looks in the wheel well of the driver's side, which has been thrust back almost into the passenger compartment. He then stands up and walks around the vehicle, nodding his head as if in agreement as his eyes dart to various areas of the body and frame. After completing a lap around the vehicle he returns to the driver's side and again places his hand on the hood. He bends down and looks in the wheel well again.

He stands up and asks if the technician who is standing behind us can open the hood. The technician opens it. Tushar looks inside the hood, running his hands along the stabilizer beam and across the cylinder head. He looks down to peer inside the wheel wells, but it seems as though he cannot see anything. He steps back from the vehicle and takes a wide look. Then he gets down on his hands and knees and cranes his neck trying to look under the front bumper. He cannot see what he wants to see so he lies down on his belly and slides his head as far under the front bumper as he can. He looks around for a minute then flips over to his back. He continues to look up inside the bumper, and he puts his hand up to feel around inside. He scoots from the middle of the vehicle toward the passenger side. He keeps feeling around until he suddenly gets a strange look on his face. He feels around some more before pulling himself out from under the vehicle.

He stands up and looks inside the hood again, running his hand between the radiator and the front bumper. He turns around to face the technician and asks him what the conduit is that he is feeling. The technician comes over to take a look and says that it is an electrical conduit for the radiator fan. Tushar smiles and turns to me saying that he doesn't have this conduit in his simulation model. He asks the technician if the conduit is supposed to be there. The technician says he will check and walks away. Tushar explains to me that if the conduit is supposed to be there, then the design engineer who owns the radiator must not have updated the conduit in his model in CAD for the files that Tushar downloaded to build his vehicle model.

He says he thinks that the absence of the conduit in his model might explain the differences between the test results and the simulation results.

After a few minutes, the technician comes back with a thick binder. He shows Tushar a page from the build log indicating that the conduit is supposed to be there per the CAD files for the radiator system pulled on a certain date. Tushar smiles widely, saying that he thinks he pulled the CAD files for the model at a date earlier than what is in the binder.

Later that afternoon, back at the design center, Tushar downloaded the CAD file for the radiator as it appeared on the date indicated in the build log at the proving ground. He found that the electrical conduit was indeed in that file, but not in the file he used to run his analysis. He substituted the new file for the old file and within ninety minutes he found that his simulation results nearly exactly matched the results obtained from the physical test conducted at the proving ground. Tushar seemed relieved.

Tushar's hands-on investigation of the physical vehicle is an example of how automotive engineers, unlike the structural and hardware engineers we studied, routinely moved back and forth between a physical world of parts and a virtual world of parts. Automotive engineers moved back and forth between these two worlds because they could not dare to consider replacing physical tests (in which humans manipulated physical equipment and physical prototypes) with computer simulations (in which manipulation was virtual) unless and until computer simulation results matched physical reality.

Although the mismatch in results between physical and virtual tests often derived from engineers' mistakes in implementation (as when Tushar employed an old version of the CAD files), and not from deficiencies inherent in the computer applications themselves, managers at IAC were concerned about this discrepancy between the physical and the virtual, and blamed the virtual. Many managers and design engineers placed their trust in parts that they could hold, see, feel, and smell; unable to similarly handle virtual parts, they did not trust the results of computer simulation. Collin, a design engineer, reflected these sentiments:

I'm not going to bet my job, and the company's wallet, and someone's life on a simulation. I mean, how do I know it was done right? How do I know it really represents what's going on? You can do anything to make a simulation work: change a number here, another number there. Besides, I go to meetings and see some analyst say he ran the test three times and got three different results. Give me the hard data from the real test and then I'll consider making a change.

This bias against simulation ran deep at IAC. Many managers had worked at the company through the 1970s, 1980s, and even the 1990s, when engineers did not routinely conduct computer simulations. In those years, engineers viewed computer applications for simulation as "prototype" technologies for experimentation in the R&D division; the application of simulation in regular engineering design work seemed a long way off. When early computer simulations made their way out of R&D and into engineering, many of those same managers saw simulation results that did not correlate well with the results of physical tests. As a result, these managers were justifiably cautious about allowing design engineers to use simulation results to inform designs. As Reggie, a design engineering manager, explained:

> My design engineers will always ask [analysis engineers] if the simulations correlated with the results. I tell them they have to. I've seen enough analyses over the last fifteen years I've been here that were wrong to be reasonably worried. I suppose we sort of have bias to real results. Once you've confirmed what your computers tell you with real tests, then we'll talk.

This ethos of distrust in the results of simulation ran so deep throughout the ranks of IAC that managers regularly admonished analysis engineers not to believe the results of simulations and to scrutinize simulations thoroughly. To reinforce this message, management required every new analysis engineer at the company to participate in a training course that covered basics on model building and analysis. A training manual accompanied the course. Figure 3.1 shows the reminder that appeared as the single entry on a sheet of paper at the beginning of each of the manual's eleven sections; the reminder prods analysis engineers to remember that computer simulation models, being virtual and not physical, were not true indicators of vehicle performance.

With so many managers and engineers skeptical of the veracity of computer simulation models, design engineers refused to give up their reliance

DON'T BELIEVE THAT
MODEL IS REALITY

Figure 3.1
Training Manual Reminder for Analysis Engineers.

on physical parts and physical tests. Design engineers like Sunil and David wanted access to physical parts so that they could verify that what they read on the screen they also felt in their hands. Analysis engineers like Rachel and Tushar wanted to be close to the proving grounds so they could witness physical tests firsthand and inspect crashed vehicles afterward. The logic of belief that valued the physical over the virtual meant that managers and engineers insisted on running an engineering process in which virtual testing did not replace physical testing but shadowed it. No one, not even analysis engineers, believed in a virtual test whose results a physical test did not confirm. Engineers allocated to computers only work that they also conducted without computers, drawing on the considerable physical resources of the proving grounds.

In short, despite strategic initiatives from company executives to move toward an allocation of work that favored the computer, the reality in automotive engineering offices and testing facilities was that the computer remained an untrusted, and hence restricted, member of the team. Although many analysis engineers, if not design engineers, shared company executives' vision of much larger role for the computer in the automotive engineering design process, they accepted that this expansion would occur in the future, not the present, and they viewed their daily efforts as striving toward that vision.

Different Logics, Different Outcomes

Faced with different logics of understanding, speed, and belief, the engineers in our study made different technology choices in the human-technology relationship. Structural engineers chose to minimize the computer's role, hardware engineers chose to maximize its role, and automotive engineers, hoping one day to maximize the computer's role, placed it in balance with the role of physical parts and physical tests. These different choices severely undermine explanations of technology choices that arise from existing perspectives.

Technological determinist arguments fail to adequately explain our findings because, although each occupation had available to it the same types of advanced information and communication technologies for engineering work, including sophisticated computational, logic, and graphical computer applications, computers did not prove indispensable for engineering work in all occupations. In the language of Fitts lists, there was no clear fit between computer applications and the tasks of design and analysis such that computers were a natural choice for all the engineers involved.

Social constructivist arguments fail to explain our findings because the computer's role and patterns of use did not vary by the firms in our study. Let us consider the two extremes in the computer's role that we uncovered. Engineers in all three structural engineering firms in our study (Seismic Specialists, Tall Steel, and Tilt-Up Designs) opted to minimize the computer's role. Their peers in all three hardware engineering firms that we studied (AppCore, Configurable Solutions, and Programmable Devices) opted to maximize the computer's role. All the structural engineers refrained from allocating even the most routine of tasks, such as checking shop drawings, to the computer. All the hardware engineers assigned to their computers every task they could. These patterns of use did not vary by firm. Identical choices across firms within an occupation suggest that local, contextual factors were not instrumental in shaping these choices.

Materialistic voluntarist arguments such as sociomateriality and critical realism are also problematic in our case as explanations of technology choice. These perspectives would suggest that differences in the types of tasks to which engineers assigned their computers would be related to the ways that the material features of those computers (and the software that ran on them) would merge with the particularities of the organizational contexts, producing unique technology choices across organizations. But we see that technology choices were, in fact, similar across organizations (or groups at IAC) within the same occupation. We cannot attribute similarities in technology choice, which occurred within but not across occupations, to materialistic voluntarism because the technologies used across the organizations were quite similar. In hardware engineering and automotive engineering in particular, the engineers' computers were similar in terms of speed, memory, and processing power. (And, arguably, had structural engineers needed fast, powerful computers, they would have purchased them; all engineers, in short, had such computers available to them.) Moreover, the structural engineers and automotive engineers employed similar CAD software in design and simulation technologies in analysis. For these reasons, these explanations fall short.

Managers proved no stronger determinants of outcomes than did technologies, which dampens the ability of deskilling or upskilling theories to explain our findings. Although managers were key players in technology acquisition and implementation choices in each firm, those choices were not driven by a universal managerial quest for control. In fact, structural engineering managers were keen to minimize the computer's role for the very reason that they wanted to increase, not decrease, engineers' control over their work process. Likewise, hardware engineers sought more discretion over their time, not less, through their allocation of tasks to computers. The thoughts of automotive engineers seemed more attuned to issues of cost and time than they were to control.

What did shape engineers' technology choices when allocating tasks to the computer were occupational factors. These occupational factors included, among others, aspects of the markets served, the basis of competition, product liability and government regulations, the extent to which engineers could fully test the soundness of their designs, and the extent to which technologies in a given field were transparent, allowing engineers to "see" into them. Factors that were important in one occupation may have played little or no role in choices in the other two occupations. For example, structural engineers' inability to see into their analysis software applications made them reluctant to assign tasks to the computer, strongly shaping their choice to minimize the computer's role. Hardware engineers could see into their programs, often altering program code, but this ability was not what shaped their choice to maximize the computer's role; it merely facilitated it. Even with this facility, the hardware engineers could have chosen to minimize the computer's role. What forced the hardware engineers' hand was not technology transparency but the rush to gain market share under the dictates of a mass market. All told, although occupational factors strongly shaped technology choices, their sway (or lack thereof) was not uniform across occupations. Occupation mattered.

As a result, we saw differences across occupations in terms of the tasks that engineers allocated to computer applications, the extent to which engineers employed the computer, and the outcomes of engineers' varying degrees of computer use. One of the most striking outcomes that we observed was the composition of the technological world in which engineers worked.

Structural engineers, who followed a logic of understanding, minimized the computer's role in their work. Consequently, their workplace was littered with physical artifacts. Paper artifacts were especially prominent and included calculation sheets, sketches, and a host of formal drawings in a variety of sizes. Because structural engineers minimized the computer's role, they often worked with their computer screens dark or their computers turned off, the monitors shoved into the back recesses of their cubicle desk. One could observe them for hours, as we did, and not see them turn once to the computer.

The technologies that took center stage in structural engineers' world, then, were technologies connected to working by hand. These technologies included pencils of various hues, handheld calculators, liquid Wite-Out, straight edges, power erasers, and staplers. The structural engineers used these simple technologies to help them employ approximate analytic methods, often in the form of simple equations, to confirm that modeling assumptions were sound, and to create 2-D representations of buildings. Their use of the computer was limited.

Hardware engineers, pushed by a logic of speed, worked in a virtual world in which almost all of their work products were digital. As a result, their cubicles were largely devoid of the clutter of normal office supplies and engineering implements. Plastic toys and water guns were far more prevalent in their cubicles than were calculators, grid pads, or even pencils. Like their structural engineering peers, hardware engineers spoke of doing work "by hand." For hardware engineers, however, this term did not signify the removal of the computer from the allocation equation. Rather, the term simply meant that hardware engineers could not fully turn over the task to the computer by means of, say, an automated program. To do something by hand as a hardware engineer meant to work steadily in front of the monitor, typing in commands at each prompt and working slowly and interactively through a series of design steps. Having maximized the computer's role, the hardware engineers rarely worked without one; the computer was the most prominent feature of their cubicles.

If structural engineers worked in a physical world of paper and hardware engineers worked in a virtual one of digital artifacts, automotive engineers lived in a world that straddled the physical (in the form of automotive parts) and the virtual (in the form of digital simulations). Engineers spent many of their working hours in front of computer screens on which they rendered parts and performance into numbers and images. But off-screen, in engineers' cubicles or down the road at the proving grounds, lay a bevy of physical parts and vehicles available for the engineers' inspection. To accomplish their work, the automotive engineers had to move between the physical and the virtual worlds. They did so with the ever-present awareness that the physical did not lie, and the virtual sometimes did, though often not through its own fault. Even though the company executives pushed the engineers to reduce their reliance on physical parts and tests, and managers were keen to achieve cost and other savings, the automotive engineers could not abandon the physical. For them, physical parts, physical tests, and the computer remained joined in a balance that, for now, held the computer in check.

Another clear outcome of engineers' choices in the human-technology relationship was how tied they were to any given task, or, alternatively, the extent to which they could do two or more tasks simultaneously (multitask). For structural engineers, their choice to minimize the computer's role meant they were tightly tied to whatever task occupied them. Progress on that task could occur only if the engineers were directly involved in its activities. By contrast, hardware engineers typically had programs running for them in the background while they tackled other tasks in the foreground. This combination of working interactively on some tasks and pushing others to the background provided a high level of productivity for hardware engineers, which aided them in their quest for speed. Like hardware engineers, the automotive engineers in analysis had simulations running for them on a bank of servers, and their simulations took hours if not days to complete. But the analysis engineers at IAC were not as free as hardware engineers to use this time for anything other than more work. Their need to balance the physical with the virtual in part explains why they were constantly busy with one or the other. As a result, we rarely saw automotive engineers of any ilk playing games or reading casual material on the Internet while at work.

This outcome with respect to multitasking illustrates that a choice about which tasks to assign to the computer and which tasks to assign to the human had clear ramifications not just for *what* work engineers did but for *how* they did their work. In particular, this choice fashioned the tempo and cadence of engineering work. Work progressed rapidly for multitasking hardware engineers, but came with the price of having to attend to multiple events and interruptions, as when programs completed when the engineer was engrossed in writing or debugging code. To manage multiple tasks, all the hardware engineers we observed kept several windows open on their computer monitor to provide quick access to different domains. For example, one window might have held the text editor for writing code, another the email application for coordinating with colleagues, a third a directory prompt at which to issue program commands, and a fourth another directory prompt for accessing files. Some hardware engineers employed two monitors to provide yet more screen space for juggling multiple tasks.

Attending to multiple tasks served to fractionate hardware engineers' work compared to structural engineers' work. Structural engineers never had the situation of two programs running simultaneously. For most of them, their computers were not powerful enough to support simultaneous application use; we often witnessed computers crashing when three or more applications were open at the same time. Additionally, the structural engineers could not run two programs simultaneously because their analyses finished quickly. Faced with such fast completion times, the structural engineers found attending to two programs at the same time cognitively impossible. For these reasons, the structural engineers never had their work interrupted because some other task had finished in the background. Consequently, their work pace was even and slow, representing as it did the speed of human problem solving.

With large simulations running in the background, the automotive analysis engineers had time for long, uninterrupted stretches of work, punctuated by the completion—after hours or days—of a simulation, or marked by a trip out to the proving grounds some forty miles away. Compared to structural and hardware engineers, the automotive engineers also attended more meetings to coordinate work across the many groups involved in vehicle development, which took them away from both their computer and their physical automotive parts. The tempo and cadence of their work were thus characterized by major shifts in activity that found them in front of the computer, at the proving grounds, or in a conference room. This pattern of work fell somewhere between the slow (human) pace of structural engineering work and the fast (computer) pace of hardware engineering work.

Overall, the structural engineers and hardware engineers we observed seemed at peace with their technology choices in terms of the allocation of tasks. The structural engineers were comfortable in limiting what work they allocated to the computer, although younger engineers did pine on occasion for more computer analysis work, similar to what they had experienced in their school days. They accepted, though, that the work demanded their attention. The hardware engineers were equally comfortable in allocating as much work as they could to the computer; none of them yearned for the days when the work was more manual, and they rallied around new technology purchases. The automotive engineers we observed, however, had yet to reach an easy peace with their technology choices. They wanted to abandon the physical, but they were never sure if and when they could believe that the virtual faithfully represented reality. Consequently, they straddled the realms of the physical and the virtual daily as they worked to discern where the truth about the performance of their vehicles lay.

4 To Automate or Not?

Few of us use but a single technology when we work. When we write a report in a word processing application, for example, we may insert into the report a table that we built in a spreadsheet application or an image that we modified in a graphics application. The insertion of the table or the image requires the applications to work in an integrated fashion, allowing the output of one technology (say, the spreadsheet application) to serve as the input for another (here, the word processing application). Engineering work is rife with examples of technologies working interdependently in this manner, with work artifacts often traversing a long line of technologies as engineers complete successive tasks of design and analysis.

One would be hard-pressed to find discussion of this interdependence among technologies in the organization studies literature. Although task interdependence—the extent to which individuals must rely on one another to accomplish their work—has long drawn the attention of organizational scholars,[1] technology interdependence, by which we mean technologies' interaction with and dependence on one another in the course of work execution, crops up only in the literature of production and operations management, which is concerned with the efficient running of manufacturing systems. As factory automation in manufacturing systems grew in the form of linked assembly lines, numerically controlled machines, robotics, and computerized equipment, so did technology interdependence. This interdependence among technologies has come to affect production outcomes so strongly that ensuring that the output from one technology can be effortlessly employed as input for the next one is essential today for success in countless manufacturing settings.[2]

Advanced information and communication technologies are becoming as prevalent in knowledge and service work as they are in factory work, primarily in the form of software applications. With the ubiquity of computer technologies in these white-collar fields there has come a surge in technology interdependence, which, just as in the case of production systems, must be managed well to ensure the successful completion of work. In this chapter, we investigate how the engineers in the three occupations we studied managed technology interdependence, or the technology-technology relationship. In particular, we are curious to understand when and to what extent engineers streamlined the flow of work across technologies through automation, and why, at times, they opted not to do so.

Overall, we found that when making choices with respect to the technology-technology relationship, the structural engineers were motivated by a desire to avoid errors, and hence shunned automated links between technologies. The hardware engineers' motivation was twofold: they desired to free up time for the writing and debugging of code, and they had a strong desire to speed up work. For these reasons, they embraced automated links. The automotive engineers' motivation was even more complicated than that of the hardware engineers; it reflected mixed desires across groups to improve quality, or to increase the speed of product design, or to standardize practice. These mixed desires resulted in mixed choices with respect to automated links.

To facilitate an understanding of how and why the engineers made the choices they did, we first establish some terms and concepts that will provide a means to assess technology interdependence and analyze automation.[3]

Assessing Technology Interdependence and Analyzing Automation

Organizational scholars have typically measured *task* interdependence by asking workers to rate on anchored scales (e.g., "1 = not at all, 5 = very much so") statements such as "I have to work closely with my team members to do my work properly."[4] Because technologies cannot answer such questions, we require a different approach to explore *technology* interdependence. We begin by tracing how the engineers we studied linked their technologies. Specifically, we consider whether and how engineers used the output of one technology as the input for the next.

We developed the concept of a *technology gap* to enable this tracing by describing the "space" between two interdependent technologies. The existence of a technology gap signals (1) that two distinct technologies exist and (2) that if a work artifact is to be processed first using one technology and then the other, it must travel between them. A technology gap may span a physical distance, as when users transfer digital files from one computer to another by loading them on a USB stick from the computer, then walking down the hallway and dropping the stick on a colleague's desk for her to upload the files on her computer. Alternatively, a technology gap may represent a virtual distance, as when users take an image from a graphics application and drag it across the screen (or copy and paste it) to insert it into a report in a word processing application.

An example of a technology gap among the engineers we studied was that between CAD software, which the structural engineers used to draw their design solutions for buildings, and analysis software, which they used to test their building designs. The engineers would import into the analysis software drawings in CAD files with the intent of using the drawings as the base for models of buildings whose performance under various conditions (e.g., gravity loading or wind loading) they would test in the analysis software.

If an engineer were to accomplish her work perfectly in a sequence of technologies as she completed one task and began the next, we would expect a linear progression, or *forward workflow*, of her product as she encountered technology gaps. For example, as the path of black arrows depicts in figure 4.1, when a hardware engineer finished writing her Verilog component code in a text editor, she might next have submitted it to a lint tool to ensure that it was syntactically correct before directing it to a testbench creator to produce compiled code, as well as RTL hardware and tests. The engineer would next have submitted the compiled code, RTL hardware, and tests to a simulator for testing, possibly ending with a plotter to produce graphs of the results.

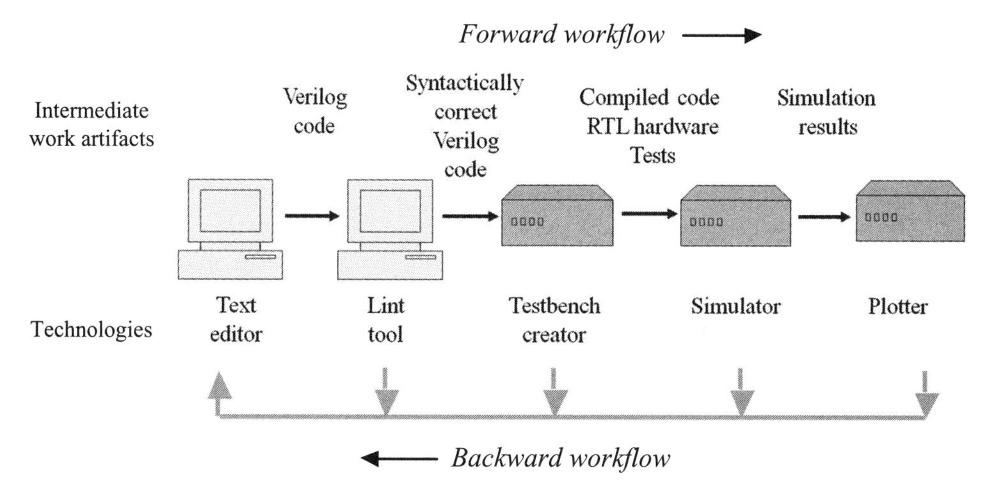

Figure 4.1
Technology Interdependence in Hardware Engineering.

Engineering work, however, rarely proceeds perfectly. Mistakes arise, causing the product to be sent back to a previous technology to be reworked in an attempt to correct the flaw. In the case of the hardware engineer and her component code, the simulator may have detected problems in the code and yielded feedback, perhaps indicating the location of error or the code elements involved in the error. The engineer would then employ that feedback as she modified her code in the text editor. Any technology in the path of the hardware engineer's workflow might likewise identify problems, causing the engineer to rework the code back in the text editor. Reworking in this manner gives birth to *backward workflow* across technology gaps, as the gray path in figure 4.1 indicates.

The existence of backward and forward workflow means that an engineer might face a gap in moving from one technology to the next, in accordance with forward workflow, or, alternatively, in moving from one technology to an earlier technology, in accordance with backward workflow. Thus, we noted whether an engineer who encountered a technology gap intended to traverse it according to a *forward* or *backward workflow*. The difference was important because it allowed the possibility that engineers may have treated differently gaps arising in the course of rework, or faulty design, from gaps arising in the course of a first or altered design. In other words, paying attention to gaps in the backward workflow as distinct from the forward workflow helped us gauge how comfortable engineers were in allowing automation to handle mistakes.

Finally, our fieldwork made clear that some gaps are more difficult to traverse than others. Returning to our example of a report in a word processing application, if we wish to place a digital photograph in the report, we may need to first transfer it from a camera to a computer's hard drive, which requires finding the camera's USB cord, attaching one end of the cord to the camera and the other end to the computer, initiating the commands to launch a transfer, and then waiting for the transfer to execute. Next, we would need to locate the transferred photograph in the hard drive directory and then insert the photograph into the report, perhaps by copying and pasting it. By contrast, inserting a spreadsheet table into our report might be easier than inserting a digital photograph if we had created the table on the same computer we used to create the report: We would need only to open the spreadsheet application and then copy and paste the table into the report. To capture this difference in effort, we conceptualized gaps as having a *width*, where width serves as a measure of difficulty in gap traversal. We determined gap width as a function of the number or complexity of steps involved in traversing a gap, as well as the time required for gap traversal.

Among the engineering processes we observed, we called a gap *wide* if its traversal required a significant amount of time (as much as an entire afternoon) and possibly multiple actions (e.g., manually transferring data points one by one across several input screens). A wide gap arose one day when Cindy, a structural engineer at Tall Steel, intended to analyze in an application called Ramsbeam the deflection of a set of steel beams she had designed in AutoCAD software. Without an automated means to transfer data from AutoCAD to Ramsbeam, Cindy had to visually read the numbers from the AutoCAD screen and then manually type them into the Ramsbeam interface. Making matters worse, the transfer involved additional steps of calculation to transform the output into appropriate input because the AutoCAD numbers were in different units than the ones Ramsbeam would accept. As Cindy explained,

> I get the dimensions from CAD [*she points to the text lines at the bottom of her screen, where numbers appear*], but Ramsbeam is in kips per feet and CAD is in pounds per square foot, so I use my calculator to do the conversion. [*She picks up her pocket calculator and enters some numbers, then types the result into the entry field of the Ramsbeam window.*][5]

We called a gap *narrow* if the output of the first technology became the input to the second one in a very short time (as little as a few seconds, and often in less than a minute) with few related actions (often simply typing a single command). In an example of a narrow gap, Eric from Configurable Solutions created a diagnostic test (named "DIAG-103") and was ready to run it against a microprocessor configuration. To do so, he simply typed a command, as described in our field notes: "At the prompt in the lower left window, he types: make vcs.log DIAG-103." Eric's gap traversal took much less time and effort than did Cindy's, reflecting the difference in gap widths.

Automation closes technology gaps, and thus by definition reduces gap width to zero, eliminating the need for human effort to turn output into input. Sometimes automation may arise when technology developers internalize within one technology a function previously performed by means of another technology. Spell checkers illustrate this point. In 1979, MicroPro International released the word processing application WordStar with the stand-alone spell checker SpellStar.[6] To check the spelling in a document created in WordStar, the user had to temporarily suspend WordStar to open the document in SpellStar and then run the spelling check. The technology gap between WordStar and SpellStar disappeared in 1985 when spell checking became a feature in WordStar. Checking spelling no longer required launching a separate application; rather, the transfer to this function was automatic.

In other cases, automation may arise when technology users develop shortcuts, as when they write software scripts to run multiple technologies with a single call from the prompt. In our study, a hardware engineer who was confident that his code was error-free may have wished to run his code through the lint tool, the testbench creator, and the simulator all in one stroke. To do so, he may have written a short script to call each technology in sequence and to feed to each technology the output of the one before it, freeing him from having to be on hand to initiate each technology call. That was essentially what we saw Raymond at Programmable Devices doing in chapter 3 when he built an interface to run a series of commands.

With these three concepts in hand—technology gap, workflow direction, and gap width—our objective becomes understanding what kinds of gaps engineers encountered and why the gaps existed. A predominance of wide gaps, for example, suggests that an occupation has yet to automate much of its work transfer across technologies and would steer us into exploring why the occupation had not automated its work processes. Conversely, the prevalence of narrow gaps might suggest a push toward automation and would prompt us to investigate the motivation for decreasing the time and effort required to traverse technology gaps.

What we found across the three engineering occupations we studied were very different approaches to automation and to managing technology interdependence, with reasons for each approach related to occupational context. We start with the parallel cases of structural engineers and hardware engineers. Their cases are parallel because, as we discussed in chapter 2, the managers in these two occupations divided engineering labor such that engineers within the same group performed the same tasks as their colleagues, but on different components of the product. In other words, the managers divided the product into components, with each component requiring the same processing steps or tasks. The managers then assigned the components to engineers, who completed all the tasks on their respective components.

Figure 4.2 illustrates this division of labor in the case of three engineers. In the schematic, managers assign each engineer a unique component (components i through iii, respectively), on which the engineers carry out tasks 1 through 4 using identical technologies A through D. In the case of structural engineering, we might consider the product in figure 4.2 to be

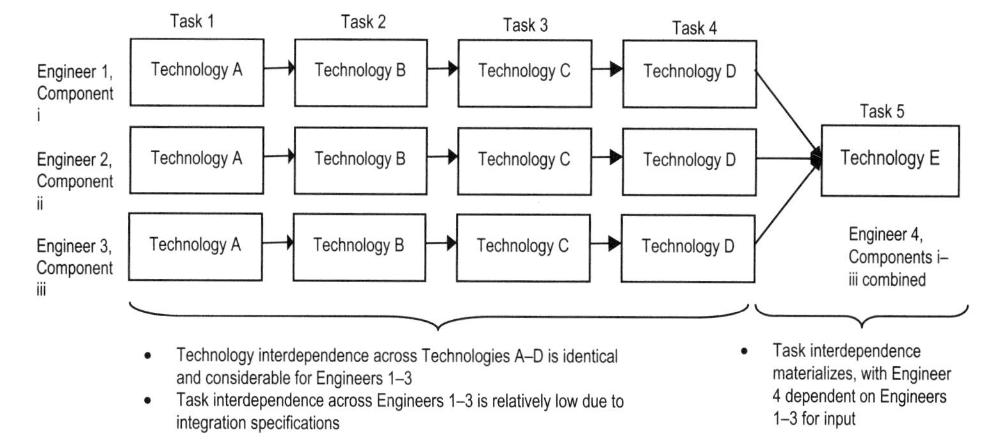

Figure 4.2
The Case of Structural and Hardware Engineers: Technology Interdependence and Task Interdependence Disentangled.

a three-story building, with each engineer designing a separate story and conducting analyses of it. When all the steps are completed for all the components (here, stories), a single engineer (engineer 4 in figure 4.2) assembles the components into a whole product and carries out whatever remaining design and analyses are required (task 5, via technology E, in figure 4.2).

Under this division of labor in structural and hardware engineering, each engineer employs the same technologies as his peers but uses these technologies to process a unique and distinct part of the final product. As a result, all engineers within a group experience technology interdependence in the same way. That is to say, because the engineers within a group use the same technologies to perform the same tasks (in figure 4.2, technologies A through D, in sequence for tasks 1 through 4), they encounter the same technology gaps. This situation was true of the structural and hardware engineers we observed.

As figure 4.2 indicates, technology interdependence does not match up neatly with task interdependence in this division of labor. Because of a priori norms and specifications that dictate how components are to be integrated into a whole at the end of the process, engineers under this division of labor can work without continual and minute coordination with their colleagues while designing their part of the product. The final integration of all the components into a whole occurs after each engineer has dealt with technology interdependence across a long sequence of operations. Only at this point does task interdependence with respect to making the current product materialize in any significant way, with engineer 4 dependent on engineers 1 through 3 for direct input (figure 4.2).

The technology interdependence experienced by the automotive engineers we observed was different from that experienced by the structural and hardware engineers because the division of labor was much more fractionated in automotive engineering. As noted in chapter 2, the automotive engineering managers divided up among engineers not just the product but also the tasks for each part of the product. As each vehicle component made its way through a series of tasks, it traversed a series of handoffs among engineers. Consequently, the experience of technology interdependence varied across the three engineering occupations we studied because the engineers themselves faced distinct sets of technology gaps, which were often unique to the occupation.

Figure 4.3 illustrates how technology interdependence was manifested among the automotive engineers in our study. In the figure, engineer 1, who performs task 1 on component i, faces different technology gaps than engineer 2, who performs tasks 2 and 3 on component i and who faces gaps different still from those faced by engineers 4 and 5, who respectively perform tasks 4 and 5 on component i. Moreover, technology interdependence and task interdependence intertwine repeatedly because handoffs

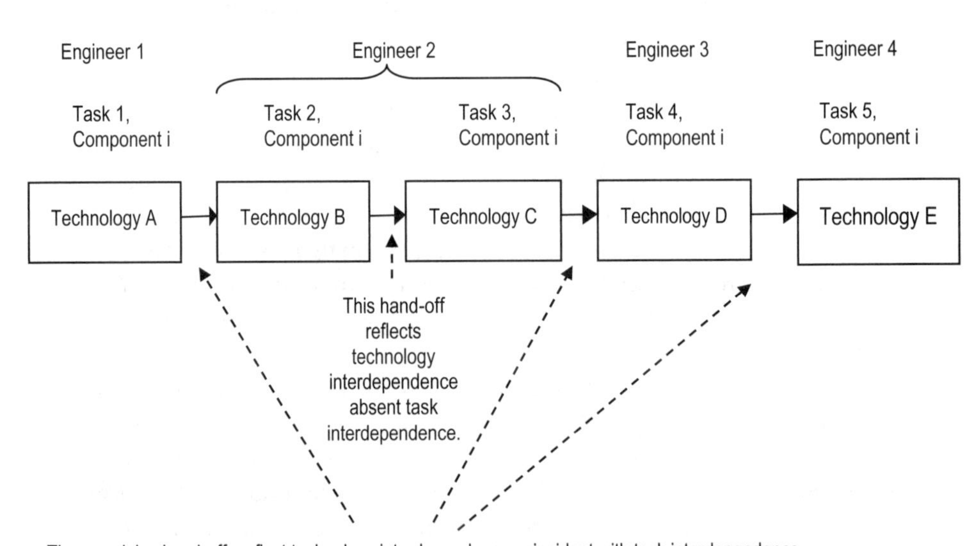

Figure 4.3
The Case of Automotive Engineers: Technology Interdependence and Task Interdependence Entangled.

from engineer to engineer often occur coincident with handoffs from technology to technology (as figure 4.3 depicts in every handoff except the one between technologies B and C).[7] This graphic illustrates what we observed among the automotive engineers.

One immediate outcome of this division of labor in the case of automotive engineering was that any decision to automate the transfer of output to input across a gap between two technologies required a decision as to which engineer should perform the newly combined operation, the engineer employing the current output technology or the one employing the current input technology. In the case of structural and hardware engineering, decisions about how to automate transfers across technology gaps only rarely involved considering task interdependence. Thus, differences in how managers divided engineering labor shaped how engineers experienced and managed technology interdependence.

In light of this observation, we might reasonably anticipate that automation would be more easily achieved, and hence more prevalent, in structural and hardware engineering than in automotive engineering because choosing to automate when technology interdependence and task interdependence are disentangled would not invoke a broader consideration of how to divide engineering labor. But beyond this work organization factor, a number of other occupational factors shaped engineers' experience of technology interdependence, and with arguably greater effect. Thus, we found that, despite similarities in task interdependence, automation was rampant in hardware engineering but rare in structural engineering. We tease out these occupational factors to explain why this difference exists. We then turn to automotive engineering, where technology interdependence and task interdependence were highly intertwined, to understand what happens when managers attempt automation around a particular technology.

Technology Interdependence and Task Interdependence Disentangled: The Case of Structural and Hardware Engineers

In both structural and hardware engineering, 80 percent of the technology gaps that the engineers encountered were in the forward workflow, with the remainder in the backward workflow. Where we saw a striking difference across occupations was in the percentage of narrow versus wide gaps. In structural engineering, fully 93 percent of the gaps we observed were wide, whereas in hardware engineering, fewer than half (42 percent) were wide. When we separated gaps by direction, we found that in both occupations, all gaps in the backward workflow were wide. The reason for wide backward gaps was the same for both occupations: rework required engineers to stop and think about what had to be done to rectify a mistake, undo poor work, and prepare the conditions for good work. Few technologies were good enough to detect a problem and fix it without human intervention, and few engineers were willing to allot that autonomy to their workplace technologies.

What about the remaining wide gaps, which constituted 91 percent of all forward workflow gaps in structural engineering but only 31 percent in hardware engineering? The predominance of narrow, not wide, gaps in the forward workflow among the hardware engineers suggests that, compared to the structural engineers, these engineers transitioned work from one technology to the next with ease in the forward workflow. Our observations confirmed this fact and provided ample evidence that the hardware engineers embraced automation across their workplace technologies, while the structural engineers shunned it. Why this difference? We approach this question by first considering what occupational factors allowed the hardware engineers to narrow gaps, and often automate transfer, between their technologies. We then turn to structural engineering to see whether the same or different factors existed there.

Why the Hardware Engineers Embraced Automated Links

The cost of technologies is a reasonable place to begin when investigating choices to automate work. In hardware engineering, technologies are expensive, with each technology costing upward of $1 million. The high cost of technologies works against the possibility of having multiple equivalent technologies to perform any given task. With only one technology on each side of a technology gap, narrowing gaps or automating them out of existence was a straightforward if sometimes difficult task for the hardware engineers we observed because the interface problem for inputs and outputs was minimal. With the output from a single technology serving as the sole input to the next one, the engineers had to navigate only one interface at each technology gap.

These interfaces greatly attracted the attention of the hardware engineers. The engineers explicitly recognized, frequently discussed, and overtly managed the technology interdependence that the interface problem represented, going so far as to give it a name, "the design flow." We observed them arguing about whether or not to change the design flow to satisfy customer requests and mulling over how to build new design flows for new microprocessor cores. Although high technology costs' reduction of the interface problem made it conceivable to streamline the design flow through automation, high costs did not force that choice.

Instead, the engineers' motivation for embracing automated links arose from their desire to free up their time to write and debug code. Automation would provide them with more time for these tasks because it would reduce how frequently they needed to type commands to launch a set of tasks across their technology suite. Gaining more time for coding tasks proved a strong motivation for the engineers because longer blocks of time would enhance their ability to think through problems when writing or debugging code, the two tasks they did not assign to computers, as we saw in chapter 3.

Creating large blocks of time for the hardware engineers to work in the foreground by streamlining tasks in the background further aided the engineers in their overall drive for speed and efficiency. Removing the human from handoffs between technologies meant that simulations would run faster and the engineers would receive results sooner. In theory, this increase in speed along the design flow would translate into shorter product design cycles and quicker time to market. Thus, the hardware engineers were motivated for financial as well as work style reasons to speed up the design flow.

Design flow considerations often began with the purchase of a new technology. A primary criterion in new technology selection among the hardware engineers was the technology's "fit" with the design flow, that is, how easily the engineers could insert it seamlessly between two existing technologies without wreaking havoc on inputs and outputs. Such a placement would allow the possibility of an automated link. To learn about new technologies, the hardware engineers visited vendors' booths at the industry's annual conference. We saw them preparing for these visits by discussing with their colleagues which vendors to see and what technology features to inspect. Vendors pitched new technologies, as we observed during an in-house presentation at Configurable Solutions, by explaining how easily the technologies could fit in a firm's design flow. In the excerpt below from that presentation, a salesperson from a vendor pitched a technology for synthesis to the hardware engineers:

> We have Synthesizer HL for the folks that are looking to do high-level synthesis. This is a fabulous product. We like to call it "the design compiler that works." If you guys are familiar with Synthesizer I, this is our first-level synthesis product; this came out about two years ago, and it basically translates RTL-level C down to Verilog or VHDL. And it gives, finally, a path in C-base design into an RTL path. Because after all, the world eventually has to go to RTL. With that, we have syntax and some specialized

Syn GDB; basically it's open source debugger tools that you can use within your design flow.

After hearing such pitches, engineers independently evaluated new technologies' fit prior to purchase, as Abrahem, a hardware engineer at Programmable Devices, described to our observer:

I have been testing the flow of a CAD tool called "Blast Fusion" from Magma [the vendor]. So far I have mostly verified that the flow works and everything is fine, so the other projects can go ahead and use that tool.

Following technology selection, the hardware engineers often eliminated forward gaps by writing scripts that automated the transfer of output. To automate this transfer was, in a sense, to abdicate control by forgoing the opportunity to review output. The hardware engineers were comfortable in relinquishing this review opportunity because they were confident that an unsound design they created using one technology would be detected by error-checking features in subsequent technologies. The ability to fully test the soundness of designs in hardware engineering thereby facilitated the choice to embrace automated links.

The engineers' faith in their ability to fully test the soundness of their designs centered on their mastering of "coverage," the state of having conceived of all possible means of failure. In other words, the hardware engineers were confident that if they thought of every means of failure, they could test for them. Thus, the challenge was not in devising tests but in thinking of all possible means of failure. In the excerpt below from our field notes, two engineers at Configurable Solutions sat in a cubicle and discussed coverage for the task of system verification. Their immediate concern was whether their firm would purchase a new technology to help with coverage or develop its own technology:

Charles: So I'm thinking about coverage in system verification.... The question is, do you think realistically we're going to buy something to solve the coverage problem?
Enrique: We're going to write our own tool.
Charles: We are? Who's going to write it?
Enrique: We don't have any tool right now to do this, but I'm assuming we're going to have something. So either I'm going to write a bunch of printouts and write some C parameters that sort of process this thing as the worst case, or hopefully having Forte [a vendor] could be one thing I'd like to explore.... We have to have a preliminary meeting to know if we even want to talk to them [Forte]. So it's extremely far from buying it. On the other hand, in the case of Forte specifically, they wanted to do a Vera program, so there wouldn't be any money involved.[8] So basically, if technically it looks like we could use it, then probably for the next six months it would be free, so then we would have access to it.

In addition to illustrating how the hardware engineers thought about coverage and their ability to test, this conversation illustrates that the engineers' interaction with technology vendors extended beyond visiting vendors' booths at the annual industry conference and inviting vendors to give in-house presentations of new technologies. The hardware engineers also entered into partnerships with vendors for the development and licensing of new technology features. Through these partnerships, the hardware engineers gained leverage in shaping technology development. For example, they lobbied vendors to embed multiple functions into single technologies to reduce the number of technology gaps in their design flows. When talking to vendors about potential new purchases, the engineers made a point of discussing problematic interfaces and how the vendor might redesign the interfaces to fit better with the firm's flow.

Another factor that facilitated the choice to embrace automated links was technology transparency. Each technology in the design flow was transparent. Automating the links between transparent technologies did not render any technology in the flow opaque. If the engineers so desired, they could "dismantle" the links at any time to peer into the workings of individual applications. We saw engineers doing exactly that on several occasions when, in debugging code, they wanted to know at which step in the flow failure in their component had occurred. To do that, they simply overrode the automated links between technologies, manually launching each technology and examining directly its output before feeding that output as input to the next technology in the flow, and continuing in that manner until they found the source of the problem. Knowing that they could override automated links, the hardware engineers were comfortable in their choice to remove themselves from handoffs between technologies. They did not worry that automation would "black-box" their design processes by eliminating technology transparency.

Because the narrowing of technology gaps was a central concern of the hardware engineers throughout the industry, one consequence of the choice to embrace automated links was that engineers shared technology information across firms. The hardware engineers whom we studied talked with peers in other firms via online bulletin boards hosted on independent hardware engineering community websites such as EDACafé and DeepChip.[9] Through such electronic media, engineers discussed and evaluated new technologies. Information sharing among peers across firms also occurred in the pages of trade magazines, which regularly featured interviews of engineers whose firms were among the first to purchase particular technologies. The engineers in these interviews reported how well the technology fit in their firm's design flow. The engineers we observed contributed to, read, and made use of information available in print and online. For example, a reporter for an industry magazine interviewed one engineer in our study regarding his firm's use of a new logic-checking application. We observed another engineer post to a web forum about a parsing program, and a third engineer download from a vendor-supported, user-group website a script to simplify transfer to a file-sharing application.

A second consequence of this technology choice was a reinforcement of one of the conditions that made the choice possible: the limited number of technologies in the workplace for each design and analysis task. Although the primary factor that limited the number of technologies was their cost, the fact that the hardware engineers opted for a tightly linked design flow worked against the later possibility of having redundant technologies for any particular task. Because the automation of links was made easier if only one technology lined each side of a technology gap, a situation that minimized the interface problem, the existence of a link argued against increasing technology options surrounding the gap that the link had closed.

A final consequence of the choice to embrace automated links arose from the managers' support of the hardware engineers' automation and gap-narrowing efforts. As a show of this support, the managers assigned engineers to serve as "gurus" for specific technologies and charged each guru with answering colleagues' questions and keeping up to date on new features and products associated with his technology. Managers expected gurus, as part of this duty, to reduce technology gaps, as Eric from Configurable Solutions described:

> Most of these tools have ways of automating tasks and writing scripts that would help you do the common tasks. And so one of the things we have done is that the person who is an expert in that helps set up some standard scripts or templates, and in many of your applications you could just use that setup as is or do minor modifications to it, and you don't necessarily need to understand exactly every step of the way. There is an automation mechanism created by this expert.

Additionally, firms employed systems administrators to maintain the computer network, manage licenses, prepare technology budgets, and help assess the fit of potential new technologies in the design flow. Through the roles of systems administrators and gurus, managers institutionalized the push for automation in workplace technologies. In short, managing technology interdependence by automating the links between technologies was a shared goal and activity among a range of individuals in hardware engineering.

None of these efforts, however, involved backward gaps. Backward gaps in the hardware engineering flow were left wide because feedback from testing programs typically prompted code modifications. These modifications almost always required critical engineering intuition and judgment, which hardware engineers believed no technology could adequately provide. Trace logs used for debugging, for example, could highlight for the hardware engineer what kind of problem had occurred, where it had occurred, and what its ramifications were. But the trace logs could not tell the engineer what caused the problem (e.g., giving a variable a wrong name, mistakenly calculating the timing of an event, or putting the wrong data in a register). In one example of navigating a backward gap, Brian from Configurable Solutions sat in front of his computer while two colleagues, Colette and Beatrice, looked over his shoulder and helped him. The trio spent thirty-five minutes working that way, trying to debug code. When Colette and Beatrice finally left Brian to complete the debugging on his own, he explained the problem to our observer:

> So we had one bug this morning. Now we're looking at the other config space that this branch was all about, and I am seeing more errors: disagreement between our instruction set simulator [ISS] and the RTL. It's not clear yet who's at fault, RTL or the ISS. We had to make changes to both given this new behavior. Yesterday we had problems with the instruction sets, and Michael put in several fixes. Which means everyone's a suspect.
>
> So what I did was we first looked at the summary file and we're just looking at the first value. I quickly ran a "create" report, which went through and summarized all of those. The advantage there is that I can quickly go through and see which is the simplest diag that's failing.... I had to rerun it with some additional options and additional traces.... This is a wave viewer. [*He points on his monitor to the window for an application called Signal Scan and begins to describe to the observer how he is using this application not to automatically fix the bug, but to sort through the data in a meaningful way to find the bug.*]
>
> The first thing is to get basic instruction: Where in the simulation do I need to be? It says, "writes cycle 522," and we are looking for PCC address. I'm starting to zoom in, to get close. I used the time [in the simulation] to get close. I'm arranging signals by cycle and putting them in order. I mentally like to prepare, like to see where I am. Extra assurance that I'm where I want to be, warm fuzzy, at the right place.

Like the example of debugging that we saw in chapter 3, Brian's explanation of the steps he took to locate the bug in his code paints a picture of absorbed detective work in which he first had to determine how to organize and look at the results of the simulation to get a sense of where the bug might be. He thought through what was happening (the code "writes cycle 522") and knew he wanted to be near there, but not there ("we are looking for PCC address"). He ordered the signals by cycle not because he knew that doing so would lead to the bug but because doing so gave him a "warm fuzzy" sense of being in the right place. His comments reveal that the hardware engineers drew on their intuition to help them debug code; they did not know with certainty the steps to take, which if they did, might have allowed them to automate the debugging process. In short, to automate across backward gaps by entering the testing feedback as direct input into a technology for modifying code far exceeded the analytic capabilities of technologies. Realizing that contending with feedback required skilled problem solving on the part of the engineer, the hardware engineers neither narrowed nor automated backward gaps.

Why the Structural Engineers Shunned Automated Links

Unlike the hardware engineers we observed, the structural engineers paid little attention to the technology-technology relationship. They did not, for example, employ the phrase "design flow" or any equivalent term. Certainly, the structural engineers understood technology interdependence in the context of their work. They recognized the issues of transfer and transformation as work products traveled between technologies. In comparison to the hardware engineers, however, they managed this interdependence much less fastidiously and with far less desire to reduce it. In fact, motivated almost entirely by a desire to avoid errors, they shunned automated links between their technologies.

As a result, when it came time to transfer work output from one technology so that it might be the input for the next, the structural engineers were centrally involved in the transfer. Cindy from Tall Steel provided a clear example when, in transferring her AutoCAD data to Ramsbeam to create a model, she first used her calculator to convert each pounds per square foot data point in AutoCAD into a kips per square foot data point, the unit of measure in Ramsbeam. Only after making the conversion could Cindy enter the AutoCAD data as input to Ramsbeam. Without the intervention of Cindy or some other human, no transfer between these two technologies was possible.

Further evidence that the structural engineers treated technology interdependence very differently from their hardware engineering counterparts could be found in their purchasing of new technologies, which they did without the detailed probing, extended vendor interaction, and prepurchase assessment that we witnessed in hardware engineering. Perhaps as a result, new technology purchases often ended up doing little to streamline the structural engineering flow. In a case described to us by Sam at Tall Steel, streamlining failed because the modeling assumptions of a new application were not aligned with the assumptions that the engineers routinely made. Consequently, although the application, in this case, automated one task, its use required the extra step of transforming input data:

If you have a rectangular building with all the floors the same, then you ought to be able to build a model that puts in all the coordinates for you automatically, which is what this new software does. However, because some programmer assumed that your model is parallel to the *y*-axis, we have to re-input all of the data. That is a ridiculous assumption to make. Programmers don't really know what assumptions make sense.

The absence of links between technologies extended even to simple network connections among the engineers' computers. In this excerpt from our field notes in an observation at Tilt-Up Designs, we see how Sally turned to a physical solution when she was unable to access the computer of a senior engineer named Frank. This computer, in addition to being the computer on which Frank worked, was the server for the firm's project files. To save energy, Frank turned his computer off every time he left the office to visit a site or meet with a client. Consequently, Sally and the other engineers in the firm routinely found themselves without easy digital access to archived project records:

Sally tries to find the files on the computer, but gets error windows back. We head off for the storage room. We pass Frank's cube, which is right at the entry to the storage room.

Sally: [*to the observer*] Frank's computer is off. It is the server for the archives. So I could just go into his cube and turn on his computer. But if I walk all the way over here, then I might as well just look it up in the boxes. [*We go into the storage room, which is full of long metal beige racks. The racks are loaded from floor to ceiling with cardboard boxes full of records.*] The boxes here are labeled by year and the file number. I was in here recently, so I happen to know exactly where this one is.

She finds the black notebook she was looking for, looks to make sure the page she needs is there, and then says she will bring it along with her back to the cube, as she may need it again: "More than once, Frank has been gone."

Several occupational factors facilitated the structural engineers' choice not to automate the links between technology gaps in structural engineering. To begin, two factors combined to yield, over time, numerous technologies for most tasks. The first factor was the low purchase cost of software in this field. The most expensive software we encountered cost only $25,000;

most applications were on the order of a few thousand dollars. Some applications were free or nearly so: engineers brought to the workplace technologies that they had acquired in their university education, such as computer programs that their professors had written and sold to them for nominal fees of $10 or so. Inexpensive technologies meant plentiful technologies, with each new hire adding to the store and firms regularly making new purchases.

The second factor that contributed to the wealth of technologies was the enduring validity of the occupation's domain knowledge, which underpinned the algorithms embedded in computer applications. A significant portion of knowledge in structural engineering consists of general physical principles, the fundamental properties of materials, and the functioning of basic components (steel beams, concrete piles, and so forth), all of which constitute well-codified tenets or facts. As writer Mario Salvadori noted in *Why Buildings Stand Up*, many problems in structural engineering relied for their solution solely on Newton's first and third laws, plus the principle of elasticity in materials.[10] The laws and the principles of physics rarely change. The practical implication for the structural engineers is that, for example, a steel beam of a certain thickness and weight per linear foot can bear the same load today as it could a decade ago. Because the rate of knowledge change is slow in structural engineering, the computer applications of the engineers we observed, which rested on this fundamental domain knowledge, continued to yield feasible, and good, results. Consequently, the engineers had little reason to throw out old technologies, which were usually as useful as new ones. Between low technology costs and enduring knowledge, the structural engineering workplace was awash in technology options for engineers.

The task of running a gravity analysis on a structural design provides a good example. To carry out this task, an engineer could choose among at least eight technology options to aid her work: using pencil, paper, and a calculator in conjunction with a design manual to solve a series of equations "by hand"; writing her own formulas in Excel in consultation with the design manual; entering numbers into a colleague's existing template or macro in Excel; or employing one of five commercial software applications (Risa-2D, Risa-3D, RamAnalysis, Ramsbeam, or Ramsteel). Choosing which technology to employ among these many options was typically a matter of engineers' preference, with no consensus on which one was "best." On several occasions, we accurately guessed in the field the graduation period and alma mater of an engineer based on the professor-supplied software the engineer employed.

The multiplicity of technologies resulted in little standardization of input and output formats, which meant that the engineers had to resolve multiple interface problems if they were to automate transfer. For example, if the engineers could choose among four technologies for one task and only one technology for the follow-on task, then four distinct gaps existed: one between each of the four technologies for the first task and the single technology for the second task. If two technologies rather than one were available for the second task, then the number of gaps jumped to eight. Why not simply ban certain technologies to reduce the magnitude of the interface problem? The engineers presumably could have overcome this obstacle, had they so desired, by restricting the number of technologies in the workplace.

The answer to that question may be found in the inability to fully test design solutions, a factor that strongly shaped the engineers' choice. Software programs for testing and analysis could mathematically verify that a structural design was sound given its assumptions. But the programs could not validate, either empirically or conceptually, the appropriateness of the assumptions in the first place. Senior engineers, fearful of errors that might result in building collapse, cost lives, and ruin careers, stressed the value of validating their assumptions through traditional practices such as tracing load paths through drawings and using time-proven approximations to estimate forces rather than repeatedly running simulations until a "feasible" (but all too often unrealistic) solution was achieved. We recall here Sam's tale in chapter 3 of the peer review he conducted at Tall Steel in which a very skinny beam was positioned to take the full load. Sam attributed the design error to a junior engineer, who, having no way to test the solution proposed by the software, blindly followed it.

In this context, senior engineers reasoned that wide forward gaps aided in the occupational training of junior engineers by cultivating their ability to test, if incompletely, their assumptions, thereby reducing the chance of errors. For this reason, senior engineers placed little emphasis on improving work efficiency by automating forward gaps. For similar reasons, the structural engineers also left backward gaps wide. Like the hardware engineers, the structural engineers recognized that design modifications based on feedback required engineering judgment that they deemed too important to turn over to automation routines.

Liability concerns and government regulations were other occupational factors that strongly shaped the structural engineers' choice. The engineer who submitted a building's design for a county permit was held responsible should the structure fail. Avoiding problems with liability and government regulations meant designing structures free of major error. Unlike the hardware engineers, the structural engineers did not trust that software programs could adequately test for every identified type of error. Sheila from Seismic Specialists explained why she double-checked by hand the output of Risa-2D, an analysis technology:

Sheila: What Risa does, it will analyze the building ... the elements, the frame or whatever. But I still—and it does some checking of them—but I still need to do some of my own design checks.

Researcher: When you do your own design checks, how do you do that? With your calculator?

Sheila: Yeah. And looking at my ASD manual.[11]

Researcher: But not another computer program?

Sheila: No.

Researcher: Okay. Why do you do those design checks? Is it that you don't trust Risa-2D or that you're maybe not sure of your input?

Sheila: I don't think Risa.... I guess what I've been told is it doesn't have all the [state building] code checks in it. For example, we check deflection and Risa doesn't check; it'll tell you what the deflection is, but it doesn't say if it's okay or not. Also, the architect might have special requirements for deflection. They don't want things to deflect too much.

As Sheila's comments about the architect's possible special requirements suggest, the structural engineers' custom market rendered each design unique, with idiosyncratic design issues that required careful reflection and reexamination at each step of design and analysis. Unique products also demanded unique paths among the suite of technologies. For example, although engineers could record all of their designs in AutoCAD, one project might require an application for analyzing wooden structures, a second an application for analyzing steel structures, and a third an application for analyzing structures formed with concrete. In addition, although engineers could record all their designs in AutoCAD, they were likely to have carried out those designs with different design manuals, textbooks, and codebooks, depending on the materials the architect or the client had chosen. With so many different paths among technologies possible, automating links would have been difficult. In this way, the custom market in structural engineering helped shape the choice to shun automation.

The low transparency of technology in structural engineering also helped shape the choice to shun automated links. If low transparency made the structural engineers reluctant to assign a single task to a single computer program, as we noted in chapter 3, then it was not going to make them any more likely to string together several tasks across several computer programs without any human intervention. Because the engineers wanted to "see inside" every task, they were hesitant to create automated links that would decrease their already limited ability to know what the technology was doing. For the structural engineers, unlike for the hardware engineers, automating links made the black box of their technology bigger.

A final occupational factor that facilitated, but did not force, the choice to shun automated links was the structural engineers' ability to reduce product complexity. If their ability had been low instead of high, they might have relied on computers to do more single tasks, and with computers doing more single tasks, the engineers might have felt an impetus to link those tasks. However, because the structural engineers could reduce product complexity through approximate analytical methods and 2-D representations, as discussed in chapter 3, they could do many individual tasks themselves. Consequently, they experienced little pressure to automate links between technologies whose role was already limited.

That the structural engineers chose not to automate links between technologies helps explain why they had limited interaction with individuals outside the firm who might have aided them in automating transfer across technology gaps. The structural engineers did occasionally send feedback to vendors about technology problems, typically through forms on the vendors' websites. They did not, however, engage in more proactive co-development or licensing activities. Additionally, the structural engineers neither attended conferences dedicated to new technologies nor communicated electronically with other members of the profession in external firms about technologies. Moreover, they did not assign gurus to technologies. In none of the structural engineering firms that we studied was there a systems administrator. In fact, most structural engineers whom we asked were unfamiliar with the term "sys admin." Rather than hire or appoint a sys admin, engineers simply let the "geekiest" among them take charge of technologies as an extra, presumed minor, duty. In sum, the structural engineers neither created nor participated in a community dedicated to the technology-technology relationship. As a result, plying a path among interdependent technologies was largely a solitary and unassisted endeavor for the structural engineers.

Because senior engineers in particular viewed the navigation of wide gaps as beneficial for the cultivation of testing acumen and prudent in the face of liability concerns and government regulations, there was little impetus to hasten automation by limiting the number of technologies that lined each gap in structural engineering: if easy gap traversal was not the goal, then all traversal options, being rather equally arduous, were viable, leaving no good reason to deny an engineer his preference. Thus, another consequence of the structural engineers' choice to shun automated links was a multiplicity of technologies for each task in the workplace. In other words, the choice to shun automated links reinforced one of the conditions that rendered the alternative choice difficult, in that multiple technologies complicated the interface problem.

We should also mention that the structural engineers lacked one key motivation for automation that strongly prompted the hardware engineers in its pursuit: competition did not compel the structural engineers to work at breakneck speed against their competitors. Whereas the hardware engineers fought against time to gain market share, the structural engineers bid for jobs against other firms. Having won a bid, a firm no longer faced immediate competition that would prompt it to hurry the work. Rather, the structural engineers working on a project proceeded with a known schedule and deadlines. Of course, problems routinely arose, causing project deadlines to suddenly loom, and the structural engineers often spent days feeling as though they were working against the clock. In general, though, the structural engineers were not hounded by the desire for speed in the way that the hardware engineers, racing against the unknown (but feared superior) product timetables of competitors, were. Instead, the structural engineers were able to work at a pace and in a manner that allowed them to forgo automation.

Technology Interdependence and Task Interdependence Entangled: The Case of Automotive Engineers

When we first arrived on the campus of the automotive firm that we studied, we were struck by the difference in the size of the engineering operation as compared to the structural and hardware engineering firms we had studied. The structural engineering firms in our sample were professional firms, meaning that almost all the employees were engineers; these firms employed between twelve and twenty engineers, plus one or two administrative staff. The hardware engineering firms we studied had functions beyond engineering, including marketing, finance, sales, and the like. These firms employed between two hundred and two thousand people, with the hardware engineers numbering between thirty and one hundred. At its peak, in 2000, the automotive firm we studied employed 44,000 salaried workers, including more than 12,000 engineers. A full discussion of why automotive firms were large when firms in the two other occupations were small is beyond the scope of this book, but the answer has little to do with product complexity (contrary to what the managers at the automotive firm believed): each of the products designed by the engineers we studied—automobiles, buildings, and chips—were complex. All three products were intricate in their details and vast in their composition, with each composed of literally tens of thousands of components.

One aspect of product complexity, however, did differentiate these engineering occupations: the extent to which the engineers found ways to reduce or cope with complexity so that design was even possible. In structural engineering, the engineers routinely reduced complexity by taking advantage of the fact that many everyday buildings are rectangular and more or less symmetric. Buildings' rectangular shape and symmetry meant that the structural engineers could drop 45-degree lines across their plan drawings (from, say, the upper left corner down to the lower right corner of a building floor) and effectively cut their design problem in half: as a first approximation, whatever solution worked for the left-side triangle formed by the line would work for the right-side triangle as well. They also employed approximate analytical methods and 2-D modeling. In hardware engineering, the engineers could similarly reduce complexity by taking advantage of the fact that many of the components in their design were identical and required identical connections. Leveraging this sameness, the hardware engineers could work at a higher level of abstraction, replacing talk of connections with talk of information flow. Although a certain symmetry did exist in vehicles (if one side of the vehicle had two doors, the other side was likely to as well), one needed only to pop the hood to see that neither symmetry nor repetition of components was an overriding feature of most vehicles. For this reason, automotive engineers were unable to reduce product complexity for the purposes of design in the same way that the structural and hardware engineers were. Instead, as described in chapter 2, they organized the engineering workforce in a manner that was isomorphic with the product, with all the inherent complexities intact. At the very end of this division of labor around the product sat two engineering groups, design engineers and analysis engineers.

Although these two groups employed different technologies, the technology gap between them was not all that wide. The CAE preprocessor technologies that analysis engineers employed to create their models for analysis could read design engineers' master CAD files without difficulty; hence the gap between the last technology the design engineer employed (Unigraphics, a CAD application) and the first technology that the analysis engineer employed (a simulation preprocessor such as Hypermesh or Easi-Crash), although not automated out of existence, was not difficult to traverse. That is to say, transferring CAD output to preprocessor input was technically not a challenge, requiring little more than a command typed at the prompt.

The absence of automation across the technology gap between design and analysis was nonetheless not surprising because the gap coincided with a handoff from one group of engineers to another. In short, the gap was organizationally, not technically, difficult to automate; to bridge it with automation would have meant redefining the roles of entire groups. Where we might have expected to see automation (where organizationally it was easier to achieve) was across the technology gaps within a single engineering group. To explore that possibility, we centered our investigation of automation on a particular group of integrated vehicle analysis engineers, the safety and crashworthiness (S&C) engineers.[12]

The S&C group was a good choice for our investigation because all 133 engineers in this group employed the same, and rather extensive, suite of technologies to accomplish their tasks, which were similar across vehicle programs and type of vehicle (truck, car). Moreover, the S&C group was a good choice because the firm was committed to trying to reduce the cost and length of product development cycles by transitioning from physical testing to virtual testing. In this regard, the S&C group employed technologies that were critical to success in the firm's vision of increasingly virtual engineering, and hence were likely to draw improvement and upgrading attention from management. Additionally, S&C's modeling was aimed at meeting or exceeding government regulations for something that consumers, lobbyists, politicians, and manufacturers viewed as highly important: the protection of occupants and pedestrians in the event of a vehicle crash. In particular, the firm had committed itself to improving S&C's technologies, utilizing S&C's analysis results earlier in the design process (to inform, rather than simply confirm, design), and otherwise aiding S&C's performance since 1992, when the government first made public its ratings of manufacturers' performance in crash tests and, in so doing, revealed the firm's vehicles' poor crashworthiness. In short, in the minds of managers and engineers alike, the S&C group was ripe for automation.

Attempts at Automation in S&C Engineering

Whereas the motivation for automation in hardware engineering was partially a desire to free up blocks of time for coding tasks and partially a desire to speed up design, the motivation for automation in the case of the S&C group was rather complicated. In fact, by couching automation solely in terms of speed without giving voice to the other motives at play, supporters of automation ultimately doomed its acceptance among engineers.

In the string of tasks and technologies that defined analysis work, what attracted the potential for automation was the simulation preprocessor, where engineers meshed components, assembled components into vehicles, and set up models. Whose attention was attracted was more complicated in automotive engineering than in the other two occupations we studied. In the automotive firm, as among our hardware engineering firms, managers assigned individuals to aid with technology evaluation, purchase, vendor relations, and license management. But whereas much of that work in hardware engineering firms was tackled by individual engineers acting as gurus in addition to their regular tasks, aided by a handful of systems administrators, at the automotive firm, several distinct groups had formal responsibility for some aspect of technology management, and some of these groups existed solely for that purpose. These groups began with the S&C engineers and from there branched out to three official technology management groups that served the firm as a whole: the Information Systems Services (ISS) group, the Research & Development (R&D) group, and the Global Technology Production (GTP) group. Each group had its own motivation for automating the links between S&C technologies.

Within the S&C group itself, some individual engineers took the initiative, much as their hardware engineering counterparts had, to write small scripts that would automate distinct parts of the work. For example, one engineer wrote a script to automate the sending of multiple input decks at one time for simulation runs in the solver, allowing the engineers to run a batch of tests with a single command. But unlike their hardware engineering counterparts, S&C engineers almost never engaged directly with technology vendors to work hand in hand on new technology features, nor did they attend conferences for the purposes of evaluating new technologies.

These tasks lay in the domain of the ISS group, which selected, managed, and maintained technologies for all of the firm's engineers. ISS also served as a source of technical support for engineers in the case of problems with any technology. One task that ISS did not take on was the direct development of technologies, though the group often worked with vendors in that regard.

To the extent that new technology development was conducted in-house, R&D was typically at the helm of the effort. R&D engineers took as their mission the development of technologies to improve vehicles. R&D was also concerned with improving the work of the engineers who designed the vehicles, helping them to achieve better intermediate work artifacts. To this end, R&D at times developed new technologies for the engineers. Whenever R&D did so, ISS had the role of approving the technology for use, perhaps asking for modifications.

The fourth group was the GTP group, which was responsible for ensuring that the firm's engineers had the best suite of technologies available to help them achieve the strategic goals of the firm. Although ISS had to approve all new technologies, including arranging contracts with vendors and arranging for payment, the GTP group had responsibility for the larger vision of the technological landscape in engineering.

The preprocessor steps caught the attention of each of these four groups as the arena for improvement efforts. Each group, however, had a different idea about what exactly needed to be done and why. For Gene, who directed the S&C truck group, and Barry, who directed the S&C car group, the reason to focus on automation around the preprocessor was to heighten the validity of the simulation models by improving model input and setup. For Gene and Barry, getting the design engineers to accept the analysis engineers' results was the most important step in clearing obstacles to virtual engineering. Thus, the quality of the models was their utmost concern.

The first approach that Gene and Barry took to improve quality was to have senior engineers write a manual detailing how to mesh and assemble components, and to train all engineers in these methods. Based on feedback from the engineers, however, Gene and Barry soon became convinced that the manual and training sessions were insufficient. They concluded that improvement in meshing and assembly could best be accomplished by automation, by having a software program automatically implement rules and standards to mesh and assemble components. The goal of automation would be the creation of highly accurate yet computationally efficient models that would correlate strongly with the physical test results.

The GTP group was similarly interested in automating meshing and assembly tasks, a key step in moving toward completely virtual, math-based product development. But for them the issue was not one of model quality but one of speed. As Martin, a GTP engineer, remarked:

> Everything here is about speed. Simulations are only helpful if they can be done quickly. The bottom line is that if a simulation takes longer to run than a hardware [physical] test, even though it [the physical test] costs more in the short term, management will always prefer the [physical] test. That's because where you really start losing your money is when you don't get the vehicle to market soon enough. If you didn't pay, say, $300,000 for another [physical] crash test, and instead you waited an extra month for the simulation to be done, and so you delayed product launch

for 30 days, you would lose, say, something like $5,000 in profit per vehicle on like 10,000 vehicles. So that's, what, like $50 million you just lost? No one would hesitate to do a $300,000 test to save $50 million. So the bottom line is it's all about how fast you can do the math [simulation].

Martin's comments speak to how competition—the race to the market—shaped the GTP group's choice to embrace automated links. With speed in mind, the GTP group set a goal that S&C engineers should be able to create and analyze a model in two days' time, a significant decrease from the thirteen or so days that it typically took most engineers. Achieving this goal meant finding out what steps were the most time-consuming and then automating those steps. For this information, the GTP engineers turned to the S&C engineers, as Georg, an S&C engineer, recounted:

I remember my friend from GTP called me up one day and said they were working on this two-day job and wanted to know what things took the longest to do in regard to simulations. I just told him some things that were normal, like it was hard to know what the latest geometry [CAD file] was in DataExchange, and things like it took a long time to mesh and assemble the models because you had to do that all by hand. It was also the most boring.

The GTP group's concern with speed was bolstered by their awareness that the S&C engineers in some of the firm's global centers were outsourcing meshing and assembly steps to small specialty engineering firms, which could complete them much faster than the S&C engineers could. The GTP group realized that if they were to build a technology that all the firm's engineers worldwide would accept and use, they had to make sure that automation reduced the time of meshing and assembly. They began investigating a sequence of technologies that would work under an umbrella interface to automate meshing, assembly, and model setup.

The ISS group was poised to embrace such an umbrella technology. ISS struggled to maintain licenses, updates, and service for the wide variety of technologies currently available in S&C. In 1996, that variety included no less than nine different pre- and postprocessing applications, combined with nine possible solvers. The ISS group's goal was to reduce those numbers by 2003 to three and two, respectively. Although the ISS group's intention in reducing the number of technologies was to lighten their own workload, an extra benefit of that kind of reduction would be to ease automation efforts by decreasing the size of the interface problem, with fewer technologies lined up on each side of a technology gap. As with the GTP group, the ISS group's key concern was speed, but for different reasons: the ISS group focused on speed because analysis engineers across the organization were performing more and more simulations each year, straining the firm's computing power. Faster simulations would enable engineers to run more simulations without increasing the firm's computing costs. In this way, the cost of technologies shaped the ISS group's choice to embrace automated links, a solution that would force the elimination of duplicate technologies for each task.

Meanwhile, the R&D group had for some time been working on the development of a technology that would link design and analysis in the hopes of incorporating analysis results earlier into design decisions. The technology would combine features of Unigraphics, the CAD technology, with preprocessors like Hypermesh and Easi-Crash. At the same time, R&D took up the task of improving the S&C model setup. Jason, an R&D engineer who spent some time working among the engineers in the S&C truck group, observed that engineers' methods for setting up models were highly idiosyncratic. Jason began work on a technology to standardize these setup methods. He dubbed the technology "Virtual Crash Laboratory" (VCL), a moniker, he explained, that was meant to convey experimental rigor:

In a crash laboratory like the ones at the proving grounds, you have to follow a protocol and document the steps you took. You can't just say, "I'll try this and see what happens." You have to follow a procedure and show that what you did could be replicated by others who followed the same steps. That is what we were trying to do with VCL. We wanted to build the rigor and credibility of a laboratory into virtual analysis. "Virtual" does not have to mean less real, but most people think of it that way. We hoped that VCL would change that by showing if you used the tool, no matter who you were, you'd get the same results, just like if you followed the procedure in the lab.

Thus, for the R&D group, the focus of automation was neither quality (as it was for the S&C directors) nor speed (as it was for both the GTP and ISS groups, though their reasons differed). Rather, the focus was standardization of practice. Through a process of discourse and negotiation, which included an outside vendor who took over much of the development, all four groups eventually came to agree on the technology that should be developed, which in the end they dubbed simply CrashLab.

Although the four groups charged with technology management were firmly behind a single, automated solution to meshing, assembly, and model setup tasks, not all engineers were equally enthusiastic about the new technology. In particular, engineers who were led to expect that CrashLab would be faster at these tasks were frustrated when they discovered, through working with CrashLab, that it was not.

The reasons why an automated technology was slower than the existing manual methods varied, but most had to do with an incompatibility between engineers' work practices and how the technology operated. For example, CrashLab worked reasonably well for setting up an initial test. Engineers, however, often wanted to run multiple variations of a test, say by leaving all the boundary conditions and accelerometer positions the same but varying the speed of the vehicle at the moment of collision. Ordinarily, engineers made such changes quickly by altering the input deck in a text editor. In the case of CrashLab, engineers had a difficult time figuring out how to locate, interpret, and alter the information in CrashLab's input decks, which were less straightforward than the decks of the existing preprocessors. Therefore, to change a deck meant that the engineer had to reenter CrashLab and navigate part, if not all, of the graphical user interface to specify within the technology the necessary changes. Having to go through this series of steps for every variation of each test added considerable time to the engineers' tasks and significantly undermined any impression of speed or efficiency they might otherwise have formed of the technology.

Similar issues arose whenever the engineers wanted to inspect the deck to fix problems in the model, which is to say, to cross a backward technology gap. Technology transparency became an issue here because linking technologies in the case of the automotive engineers, unlike in the case of the hardware engineers, did render the resulting technology sequence a "black box" that engineers found difficult to open up. Fred, an analysis engineer, explained to us:

> Let's say you use a tool like CrashLab and it automatically generates your deck for you. Well, you don't know how it's doing it or what it's putting in there. So I guess you've lost some control over the process. This becomes a real issue if you have to debug it because it bombed out. If you didn't know what was in the deck in the first place and didn't control your output yourself, you won't know where to begin debugging it, and it could take you forever. I don't have forever. I don't even have an hour.

Worse yet, CrashLab generated output in HTML format, which follow-on plotting applications could not read as input (they could read the text-file format output of existing solvers). Thus, to plot the results of all their iterations, the engineers now had to manually enter their data into the plotter, which took time and provided an opportunity for error. CrashLab's output format was HTML because its developers wanted a clean, straightforward look that would work when presenting results to design engineers. In other words, a feature aimed at easing task interdependence between design engineers and analysis engineers worsened technology interdependence for analysis engineers.

The choice to embrace automated links by the S&C managers and various technology development and management groups at IAC resulted in continued development efforts for new technologies in meshing and assembly and coordinated action across these four entities to bring about an automated solution. These groups engaged in considerable interaction with commercial vendors at various points during the development of the solution, and each group retained its unique expertise and mission within IAC. But because S&C engineers shunned automated links through their rejection of the technology, many of the expected consequences of automated links failed to materialize. The introduction of CrashLab did not lower costs, speed up design, or improve quality. In particular, with the rejection of CrashLab, work practices did not become standard across S&C engineers, who continued to employ idiosyncratic methods across unique, independent, and different technologies.

Three Lessons We Learn about Occupational Approaches to Automation

We see in this chapter tales of automation alternatively embraced, shunned, or tried but failed. The hardware and structural engineers' choices differed—one embracing automated links, the other shunning them—even though both occupations divided engineering labor to limit task interdependence, a step that should have set the stage for automation by spacing task handoffs between engineers far apart. Automotive engineers, unable to sufficiently reduce product complexity, divided labor in a way that tightly tied task interdependence to technology interdependence. Consequently, they were left with few opportunities to automate the transfer of output to input across major engineering groups. The automotive engineers did, however, embrace the idea of automating within a group, particularly around the preprocessor. Although these efforts culminated in a new technology that linked several analysis steps and related technologies, the drive to automate fell far short of its goals because automotive engineers rejected the technology.

These different tales of automation have at least three important lessons. The first lesson is that not all engineers desire automation, the presumed pinnacle of efficiency, in their work processes. Since the days of Frederick Taylor, an engineer whose scientific management movement elevated the streamlining of operations to a quintessentially American mandate, efficiency has figured prominently in engineers' redesign of work systems.[13] Although Henry Ford claimed he owed nothing to Taylor, Taylor's efficiency ideas proved a perfect fit for the assembly line, where Ford quickly paired rationalized work with the automated transport of product between workers and machines.[14] Ever since, automation has held a prime spot in engineers' toolkits for speeding up work.

We were surprised, then, to observe senior structural engineers eschewing automation when designing the work processes of their junior colleagues. Worried that the generation of fresh engineering graduates relied too heavily on computer printouts and too little on the engineering common sense whose acquisition rampant computer use appeared to inhibit, and convinced that only careful, seasoned human judgment—not computer algorithms—could generate designs that would stand up to the demands of liability concerns and government regulations, senior structural engineers refused to automate the gaps between technologies in their design and analysis processes. Rather, they opted to force junior engineers to manually transfer (and perhaps, in so doing, reflect on) structural designs from one technology to the next through the progression of design and analysis tasks. For these engineers, automation held false promises of efficiency because streamlining operations would merely yield poorly conceived designs whose mistakes, if the engineers were lucky, county inspectors or peer reviewers would catch. If the engineers were unlucky, bad designs would slip by, possibly resulting in property damage, injury, or death, for which the structural engineers would be responsible.

By contrast, the hardware engineers and automotive engineering managers (if not automotive engineers) viewed automation as a sure path to efficiency. The hardware engineers, confident in their ability to fully test, through technology, the soundness of their designs, could automate the links between technologies without the fear of poor design that haunted the structural engineers. Automotive engineers, particularly those in the R&D group, believed that automation was the very means by which they could eliminate poor design by standardizing the modeling processes of model setup and data input. The hardware engineers and many automotive engineers believed that automation would, ultimately and safely, speed up their design work.

The second lesson we learn from these different approaches to automation is that in the two occupations that did desire automation, motivations varied. Among the engineers we observed, motivations varied because different sets of occupational factors shaped motivation. Although the desire for speed played a key role in shaping motivations in both hardware and automotive engineering, in the latter case the situation was complicated by other, competing desires. In the case of the S&C engineers whom we studied, each of the four technology management groups involved in efforts to automate around the preprocessor had its own motivation for automation, from quality to speed to cost savings to standardization of practice. Moreover, the two groups that favored speed did so for different reasons: one group wished to improve time to market, while the other hoped that speeding up simulations would remove the need to buy additional simulation technologies to handle the workload. Thus, the "why" of automation varied by group in automotive engineering, even though the "how" of automation was the same: using CrashLab to break down the work process into small tasks, standardize practice for each small task, and replace human discretion with automation wherever possible. As these actions suggest, Frederick Taylor and Henry Ford could have had no clearer offspring than the automotive engineering groups that sought to improve the efficiency of engineering design and analysis.

The third lesson that we can draw from these automation tales is that technology choices have broad ramifications that alter far more than how engineers interact with the everyday technologies they employ to do their work. For example, the structural engineers' choice to shun automated links not only shaped how engineers transferred output to input across technologies, it also defined the engineers' limited interactions with vendors, minimized the role of technology gurus, and eliminated a primary reason for dialogue and discussion among engineers across firms. In other words, in turning their backs on workplace automation, the structural engineers did not simply ensure that the age-old admonition to "sharpen your pencils" continued to have literal as well as figurative meaning for fresh engineering hires; their choice about how to turn output into input at the level of technologies was influential in shaping social interactions, or the lack thereof, across the entire industry. In a similar way, the hardware engineers' choice to embrace automated links yielded extensive vendor interaction, the roles of systems administrators and technology gurus, and industry conversations about how well individual technologies fit into design flows. In the automotive firm, efforts to automate links led to coordinated action across four engineering groups and concerted interaction with vendors. Across all three occupations, then, technology choices around automation had consequences for social aspects of work.

Notably, technology itself did not bring about these or other consequences, nor were managers universally prompted by capitalist desires to control engineers. Individual settings at each site did not lead to variety in choices about automation within each occupation: each structural engineering firm in our study chose to shun automated links, each hardware engineering firm chose to embrace them, and managers of the four groups within the automotive company we studied opted to embrace automated links (although the automotive engineers did not). Finally, it does not seem that any "sociomaterial" practice arose emergently at the intersection of technology use and social action because each occupation chose actively whether it would embrace or shun automation. Rather than uncovering support for arguments from the standpoints of existing perspectives along the determinism/voluntarism and materialism/idealism distinctions, we found that within each occupation a unique set of factors combined to shape engineers' choices.

5 People amid the Technology: Locating Engineering Work

Increasingly, technology permits work to move. Advances in information and communication technologies enable more and more people to work at a distance from their colleagues, whether they are teleworking from home, attending a web conference while traveling for business, or working out of an office halfway around the world from the rest of their work group. An entire offshoring industry has emerged as a result, as call centers in places such as Malaysia, the Philippines, and India attest. Particularly appealing to managers are the economic benefits of shifting work to other locations, benefits that include lower wages for workers outside the United States. Is engineering work portable in the same kinds of ways that call center work is? Are engineering firms poised to take advantage of salary differentials around the globe, distributing work across great distances?

Here again, we find it is not the work, technology, or complexity of the product that drove the engineers in our study to take advantage (or not) of the affordances of advanced information and communication technologies for distributing work around the globe. Rather, occupational factors guided engineers' choices about how and to what extent to employ advanced technologies in the human-human relationship, particularly with respect to where to locate engineering work. In each occupation, a combination of factors drove a need that shaped engineers' choices. Structural engineers needed to maintain a base of domain knowledge; hardware engineers needed to coordinate. For these reasons, both of these occupations chose to reject distant work. Automotive engineers, needing to lower costs and to increase speed, chose to employ distant work.

Learning and Liability: Why Structural Engineers Worked Side by Side

In structural engineering, work was proximate, not distant. That is to say, all of the engineers in the firm worked the same hours in the same office.

When asked, the structural engineers explained their rejection of distant work in economic terms. According to this explanation, firms billed engineers' hours against bid projects. To win bids, firms needed to keep their costs low, which meant they never budgeted overtime hours. As a result, structural engineers, particularly junior ones, typically did not work at home in the evenings or on weekends. In fact, junior and mid-level engineers did not receive laptops or desktops for home use, and the firms lacked computer networks that would have allowed remote connections to servers in the office. As Sheila at Seismic Specialists noted,

> I don't work at home.... I haven't had to work overtime much since I've been there and they don't—if they don't have the money to pay you overtime, they don't want you to work overtime. So the jobs that I've been on I haven't needed to, and I don't think they've wanted me to work [at home].

This economic explanation for why structural engineers worked close together and face to face is, however, incomplete. It is incomplete because by itself, it ought to have provided strong motivation for engineering firms to offshore work to countries where engineering labor was cheap. To understand why structural engineering firms did not offshore work, then, we needed to probe deeper.

In some respects, the choice to keep structural engineering work local and centralized followed from the choice, discussed in chapter 3, to minimize the computer's role in this occupation. Because paper drawings (full of notations appended by hand) and project binders (full of calculations and sketches done by hand) retained a central role in the structural engineering design process, work artifacts in this field remained difficult to port across distance. These work artifacts were important not only in the current projects for which engineers generated them but also for future projects, for which they would serve as valuable references. Each firm in our study maintained one or more "project rooms" that served as storehouses of physical documents from past projects that engineers frequently accessed for current ones. In chapter 4, we saw Sally at Tilt-Up Designs access such a room the day she found Frank's computer (the server for archived records) turned off. When engineers worked together, as they often did, they pored over these physical artifacts, many of which were large drawings that would have been difficult, if not impossible, to view comfortably on computer screens. These large drawings, rolled out on tabletops, permitted engineers to gesture indexically while discussing and thinking through design problems and solutions with their colleagues, as when they pointed to a particular feature in a design. The drawings also served as records of ideas when engineers jotted notes on them during design sessions and as communication devices when engineers handed off these drawings with notes either to each other or to drafters.

In addition to the physicality of the work, several occupational factors beyond competition shaped the structural engineers' choice in the human-human relationship to work in close proximity. Combined, these factors reflected a need to maintain the base of fundamental domain knowledge in the structural engineering workforce. One factor that shaped the choice to reject distant work was the rate at which knowledge in the field changes. Because knowledge in this field changes slowly, structural engineers were able to accrue knowledge that remained valuable over the course of their careers. Senior engineers passed down their accrued knowledge by teaching their junior colleagues, much in the manner of masters guiding apprentices. Teaching and learning were extensive because each engineer needed to gain the full expanse of the field's fundamental domain knowledge. A wide individual knowledge base was necessary because the custom market in this occupation led to individualized products for particular clients, each of whom could choose from among a variety of possible designs (e.g., a choice of wood, steel, or concrete). Engineers could not easily specialize in a particular design, especially in small firms. Consequently, each engineer needed to know how to do every kind of design, as well as how to work with every kind of element. For this reason, teaching and learning pervaded the structural engineering workplace.

This teaching and learning demanded face-to-face interaction. Junior engineers often sought to learn in face-to-face interaction with their nearby colleagues. For example, a junior engineer might have approached a senior engineer and asked for help in solving a problem that involved an unfamiliar design solution. When we asked Sheila from Seismic Specialists to compare the problems she had solved at school with the problems she faced at work, she explained what prompted her to ask a colleague to teach her:

> I can compare them [the problems], but they're a lot different. I mean, in school you just have time to do everything exactly and you have deadlines, but you always have enough time to get your homework done and there's always a professor to go to with the answer. At work there's a lot more; you have to make a lot of assumptions. There are just so many more unknowns, and you can't do everything exactly. At least the way we're set up now. At school also usually there's an example to follow that's really similar to what you're doing or you'll have learned it. For me, I have to figure out what I need to figure out. Part of my task is to figure out who to ask to learn how to do it.

Just as often, engineers voluntarily and without solicitation offered learning opportunities to their colleagues.[1] For example, in the course of examining together the day's work by looking at sets of drawings and calculations, a senior engineer might have determined on his own that a subordinate required a lesson in some general principle of design. In the following example, Cindy, a mid-level engineer at Tall Steel, had called over her cubicle wall to Sam, a senior engineer, to discuss a discrepancy between the size of a building's braces and the columns to which they were attached. Suspecting from her description that Cindy had a conceptual misunderstanding, Sam did not directly address the situation. Instead, in an effort to convey a larger point about how to view designs that featured a single horizontal element in some section, he began to quiz her about the struts that spanned the columns:

Sam: [*from his cubicle*] Where's your gravity load on the end of the strut? [*Sam comes over to Cindy's cube and stands behind her, looking at her computer screen as she sits below him. A software package for analysis is open on the screen; it displays building diagrams with output data.*] Which way did you apply the seismic load? [*Cindy points beside a diagram to indicate the direction.*] Did you run gravity only?
Cindy: No. [*She begins to navigate to a screen that will allow her to run an analysis.*]
Sam: [*Showing irritation in his voice, he stops her.*] Let's look at the model first before you run the analysis. [*Sam leans over and points on the screen to a beam in the lower left corner of a diagram and begins to quiz her.*] Is this in compression or tension?
Cindy: [*hesitantly*] Tension?
Sam: [*signaling a wrong answer and suggesting that she think again*] Under gravity loads?
Cindy: Oh, compression, I'm sorry.
Sam: So what's that one? [*Sam points to the upper left horizontal beam in the diagram and continues the quiz.*] Compression or tension?
Cindy: [*again hesitantly*] Tension?
Sam: Yeah, now is this the same load case? [*Sam points to another set of beams on the opposite side of the diagram and offers a clue.*] The wind controls in this case.
Cindy: [*explaining why the diagram cannot be used for this quiz because the numbers on the screen combine gravity and lateral loads*] I'm showing the load combination.
Sam: Show just the gravity. [*Cindy alters data in a table. As a result, the load numbers for some of the beams change on the screen. Sam points to the upper right corner.*] So what's this? [*He directs her attention to the junction of the strut and the column and finally tells her that she could have passed his quiz by recognizing a fundamental feature of her structure.*] Are you visualizing the equilibrium of this point? This is the only horizontal component so it is in compression. Okay?
Cindy: Yes.

As the example of Cindy and Sam suggests, senior engineers were the most prevalent teachers in the structural engineering workplace. Junior engineers were the most frequent learners, followed by mid-level engineers. The bulk of this in situ learning concerned technical concepts, such as compression, tension, and, more generally, how building components acted in solutions, as well as a variety of other definitions, theories, and relationships from science, engineering, and mathematical domains. Much of the remaining teaching concerned procedures (e.g., how to submit drawings, what kinds of sketches to make, how to respond to a contractor's query) and politics (e.g., how to negotiate with an architect).

Seniors mostly taught their junior and mid-level colleagues by collaborating with them, working through design problems side by side as they stood together over full-size drawings. In this manner, senior engineers guided others in discussions of design issues. These design issues included how the engineers should deal with parts of the design that were tricky theoretically (because the path that loads would take through the building was unclear), tricky materially (because the components' size exceeded the space available for them), or tricky politically (because the best design structurally was aesthetically or economically unappealing to the owner). By collaborating with their junior colleagues, senior engineers made transparent to them their own thought processes and provided templates for problem solving.

We saw this kind of collaboration in chapter 2, for example, when Rush and Tim at Seismic Specialists worked together to design piers to be set into rock. Rush's remarking to Tim that the results of their model were "way out of the game here," his double-checking his own calculations and finding and fixing a mistake (which was forgetting to multiply by pi), and his rejection of several potential solutions all spoke to how Rush provided Tim with a chance to learn how to go about solving a foundation design problem. In doing so, Rush never put the full weight of the problem on Tim's shoulders. Rush's behavior signaled, and Tim's behavior implied he understood, that they were in it together.

Senior engineers also taught by instruction, which is to say, rather than working through a problem with an engineer, as Rush did with Tim on the piers, senior engineers at times simply provided their juniors with a list of steps to follow independently. In this example, George, a senior engineer at Tilt-Up Designs, instructed Dennis, a junior engineer, on how to prepare two foundation plans, one with caissons and the other with piles, both of which were structures for transferring building loads into the ground:

George: We probably want to say, "Use these pile caps and throw in this amount of rebar as baseline," based on which one is cheaper, or meets all their scheduling problems. In general, that will determine which one they select. When one is selected, we'll begin the task of changing the drawings and doing the real pile caps, because three feet diameter concrete caissons is going to be a different spacing [than the drawings currently show]....

Dennis: [*who has never prepared such a summary before*] So base the steel on—?

George: [*correcting him*] Base the *rebar!*

Dennis: Yeah, the rebar.

George: Base the rebar just on a wild-assed guess.

Dennis: [*laughs*] Okay!

George: [*more seriously*] Based on previous jobs that had pile caps. And that's just a—

Dennis: [*interrupting and guessing the end of the sentence.*] Just a general note?

George: It might be an 8½ by 11 [inch] sketch that goes in with the bids. If we can describe it in words, that's even better. If it [the rebar estimate] gets sort of detailed, then we probably need a sketch. To get accurate numbers on these things, you would need to go through a full design and detail them all out.

Another common method of teaching was tutoring, in which the senior engineer adopted a pedagogical attitude. In practice, tutoring meant that teachers digressed from the particulars of the problem to extract larger lessons from the situation. When tutoring, rather than simply offering side-by-side help or a list of step-by-step instructions, senior engineers often quizzed their younger colleagues on fundamental concepts. Tutoring is what Sam did when he taught Cindy how to view the lone horizontal component in her design.

The considerable teaching in which senior engineers engaged their younger colleagues required colocation because teaching arose seamlessly in the course of everyday action. Teaching was best delivered in the moment. Moreover, senior engineers needed to judge how well their juniors were learning, and junior engineers needed to signal their understanding. The follow précis from a design meeting of four engineers at Tall Steel is illustrative. Sam, a senior engineer, directs most of his comments to Darren, a junior engineer, who is taking the lead on this project. In return, Darren provides most of the positive reassurances in response to Sam's comments ("Mmm hmm," "Right," "Okay"). When talking, Sam often points to Darren's copy of the architectural drawings, looks directly at Darren, and directs his instructions to Darren, at times ignoring Cindy's comments to focus his attention on teaching Darren. Although Darren is Sam's intended student, Cindy and Beverly, through their presence, also have an opportunity to learn, though less directly than Darren.

Darren goes into his cube, grabs a red pencil, and sharpens it in his electric sharpener. Beverly trails him with a notepad and plastic cup full of pencils and markers.

Cindy and Sam join us in the conference room, where we sit around a circular table. Each engineer has a hard-copy set of the 11″ × 17″ architectural drawings. In addition, there is a single set of the 1.5′ × 2′ structural engineering drawings. Later, when Ray, the architect, arrives, he will add some 3′ × 4′ architectural drawings to the mix. For the most part, though, the conversation centers on the smaller architectural drawings and the structural engineering drawings. Darren begins the conversation by filling in Sam on the situation with a porch on the side of the building.

Sam: [*looking at the drawing spread out before them on the table*] Does that mean that that whole porch is a supported slab, then? Is that what they're saying?

Cindy: [*examining the drawing*] No, it can't be. Looks like a balcony.

Sam: That's a good question, because that makes a big difference if that's a supported slab. Then we don't have retaining walls but we do have supported slabs out there. So that's, you know, two different structures. One structure, you've got, if you put a retaining wall here and backfill it. And while you're doing this stuff with Ray you need to be making up questions for the contractor because this is a construction issue, right?

Darren: Mmm hmm.

Sam: Because the contractor, because I don't know if they've got, they might be hauling dirt off the site, and if that's the case they may prefer to build a retaining wall and just throw the dirt back on over here [*he points on the drawing*], so that's a question for the contractor, which sequence does he prefer over there? [*Darren writes a note on the drawing.*] It may be that he wants a sketch from you of what that looks like in each condition. I don't know. Instead of doing, unless you can do CAD as fast as you can do a freehand sketch, which we all aspire to but none of us have ever gotten there. Um, what does this look like with a retaining wall ... with a retaining wall it is kind of an L shape put in down there, and that's another retaining wall, right? Over there [*he points on the drawing*]. So another L-shaped retaining wall is put in over there. Looks kind of, and then what is the dimension of that footing? It depends on what this is, right?

Darren: Mmm hmm.

Sam: Okay, then we can go to our retaining wall schedule to get that, our typical retaining wall schedule. Did the soils engineer give us lateral pressures here?

Darren: Um, maybe.

Sam: I think he did.

Darren: [*ready to get up to fetch the soils report*] Do you want—?

Sam: That's alright, you don't need to get them right now. I'm just saying, if you get the lateral pressures you can put that in our spreadsheet and check the dimensions. Whatever it is, we want to make sure this is minimum; we don't want to make this any bigger than it needs to be—

Darren: [*interrupting*] Alright.

Sam: —for that retention.

Darren: Yeah, I think we have all the soils report, all the tests.

Sam: And once you do that, because you make that, that would end up having to have front face reinforcement as well. Because once you tie it in with a slab at the top, all future pressures go into kind of a simple span, so that's what would be different from your schedule. You'll have the scheduled reinforcing on the back side plus you'd have front reinforcing designed for a uniform load on the back.
Darren: Okay.

They continued on in this manner for another three hours. During this time, Sam tells Darren that they need to understand what the grade is around the area of this slab. They should prepare two options for the contractor. Sam thinks they will have to build a slab with a retaining wall as one option. He warns Darren that the building will be in moist conditions up in the mountains, which means steel would not be a good choice, but a poured concrete slab would work well. Thus, the second option would be a supported slab. The difference in cost, Sam figures, should not be that great. He reminds Darren to ask Ray how tall the walls are in this area because the contractor will need to know that height for his pricing. He then tells Darren what drawings to create for this section of the building.

In this meeting, Sam taught Darren by leading him through the steps of preparing for a meeting with the architect and guiding him through the questions he should ask the architect: Is the balcony a supported slab? Will they be building a retaining wall, and if so, does the contractor want to backfill it with dirt from the site? What is the variation in grade around the slabs? In addition, Sam advised Darren how to go about his tasks, as when he told him just to freehand a sketch rather than work up a drawing on AutoCAD, that Darren should use the lateral pressure numbers from the soils engineers' report to help him determine the dimensions of the retaining walls, and that he should offer the contractor two design options: a supported slab and a slab on grade. Sam also taught Darren about design issues. For example, he told him that if they tied the slab on the grade to the retaining wall, then they would need to add front face reinforcement to the wall in addition to the back face reinforcement already scheduled because the combination of elements would act as a simple span. Similarly, Sam told Darren that the moist conditions in the mountains favored a concrete design over a steel one. Finally, Sam taught Darren how actors with whom the firm interacts thought, how Darren could aid that thinking, and how Darren ought to interact with these other actors. To this end, he explained that the contractor, when looking at the drawings and trying to determine pricing, would want to know the wall elevations and the footing.

Sam conveyed all of this knowledge—of preparation, of how to do tasks, of design, of professional conduct—to Darren while talking to him face to face over drawings that they examined together in the company of the other two engineers. Darren's constant affirmatives and his eye contact let Sam know that Darren was listening and understanding. Because all four engineers worked in the same office, they could juggle this and many similar design meetings to adjust to last-minute schedule changes, moving meetings forward or backward to allow, for example, for unexpected phone calls or tasks that took longer than expected. They had at the ready all the materials they might need to do the work, including architectural drawings, structural engineering drawings, manuals, calculators, and even a box full of bolts and other fasteners. Although advanced information and communication technologies might have made it possible for engineers to share and collaboratively examine large drawings through digital displays, the expense would have been considerable, and no system could have provided the same degree of copresent awareness that engineers enjoyed in their face-to-face setting.

The example of Sam and Darren points out another occupational factor that shaped structural engineers' choice to reject distant work: the fact that their work artifacts were not self-explanatory. That is to say, when an engineer created an artifact such as a drawing or a calculation sheet, she would, in the course of events, make quite a few decisions. For example, the engineer would decide which component to employ, what size to make it, and where to place it. In most instances the artifact recorded the engineer's final decision, but not her decision process or rationale. Why did she choose that component, why did she make it that size, why did she place it there? Engineers often could only guess at these answers when examining someone else's artifact. Sometimes the engineers struggled to glean even the most straightforward information, as when Sam began the design session by asking if the balcony in the drawing was a supported slab. Because work artifacts could not reveal all the decisions that went into them, their usefulness as templates for new projects was not guaranteed. To gain knowledge and understanding, junior engineers had to rely on senior engineers for the explanations and design rationales that artifacts did not yield.

Similarly, because structural engineers could not fully test their models either physically or virtually, they could not learn through experimentation or trial and error. In the absence of full testing capabilities, structural engineers had to rely on their assumptions. They learned to make good assumptions through day-to-day interactions with experienced colleagues. In this sense, the structural engineers' low testing capability also shaped their choice of where to locate work by placing the onus of teaching on senior engineers.

A final factor that shaped structural engineering's choice to reject distant work was liability concerns and government regulations. Liability concerns and government regulations drove senior engineers to keep work where they could see it. Senior engineers stamped with their name the final design documents for each building project; with that stamp, senior engineers became legally and professionally liable for designs. Darren from Tall Steel explained to our observer where each of his firm's engineers stood with respect to liability and professional authorization to design buildings:

This is the RFI log[2] for the 13 retail buildings that surround the hotel. Every one of those buildings is unique, not repetitive, and new problems arise all the time. Many times the answers require a sketch. Many of the sketches are related to tenant changes, because the buildings were designed before the retail tenants signed on. We fax and email the changes. Chuck and Tony are on site and they deal with it every day. [*The researcher asks who has authority to answer RFIs.*] Anyone here can send out the RFI info, but Sam is the engineer on record, so usually we will run things past him. Cindy also has her professional engineer's license. Beverly can take her exam in April and I will be able to take it next November. Tony has his PE license. The four owners can stamp drawings; they are the only ones who can take on the liability. At some point they will add Cindy and Tony to the insurance policy and then the premiums will go up. With the structural engineer's license, which is different from the PE license, you can do the highest safety buildings. A PE can do normal houses. Safety facilities are things like schools and hospitals. The PE exam is given in two eight-hour sessions, but for the SE exam, you must first work under a licensed structural engineer for five years before you can take the test.

Because senior engineers were held liable for the designs that their junior colleagues constructed, they wanted to have firsthand access to and visibility into the engineering design process. Moreover, they wanted their junior colleagues to be able to ask them questions immediately whenever problems arose. Additionally, in an effort to prevent problems, they wanted to check in with their junior colleagues on a near daily basis to see how the work was progressing. This excerpt from a quick check-in between Sally and her manager Mitch at Tilt-Up Designs illustrates how senior engineers kept tabs on their engineers while expressing their opinions and desires about how they wanted the work to be done.

Mitch arrives at his glass-walled office, which is in front of Sally's cubicle. He comes over, and Sally hands him her timesheet for him to sign it. Tia also brings hers over to be signed.

Sally: [*to Mitch*] I talked to Max Sheffield about moving the columns to the outside. He laughed. We'll move the pipes.

Mitch: You're meeting with them today?

Sally: Yes.
Mitch: Maybe next week I'll try to go.
Sally: We're moving ahead with the raised floor. One option will add Unistruts to the posts.[3] We still must address the braces. Also, here [*she points on a drawing on her desk*], we're using a pedestal.
Mitch: Big?
Sally: We could get away with a TS2×2.[4]
Mitch: How about pipe? They prefer pipe in these kinds of things. They need some adjustability.
Sally: But they could add—
Mitch: [*interrupting*] But that is not preferred. If they could just have a threaded union to adjust the pipe.
Sally: Ok, that's fine. I can use a smaller TS here, 2-P5000, a tiny pipe in between. I need to cantilever the plate above. So I was thinking....
Mitch: So that's three and a quarter?
Sally: That's a three-inch pipe.
Mitch: It fits in well. Have an external thread, with an internal thread above it that is larger. [*He pulls a red marker out of her mug and takes off its cap to demonstrate how the threads will fit together.*] You know, the larger above gives what you need—
Sally: [*interrupting*] I understand.
Mitch: If that rotates, then attach it to the Unistrut. It should attach, if anchored here and loaded here [*he points on the drawing*]. But if the base plate is welded to it, both parts don't match.
Sally: You could just adjust the bottom.
Mitch: That's true, very good.
Sally: I talked to Burt and Reggie. We discussed that maybe we could do this option in part of the area, and this other one in another part, not all the same.
Mitch: As long as the columns and braces are compatible.
Sally: Maybe since it is quicker to do this one, then come back. I said I'd come up with some sketches for them to see, give them some direction.
Mitch: Ok, good.

In this daily check-in, Mitch not only received an update from Sally about the project, he also let her know that he wanted her to use a pipe with a threaded union instead of a wooden pedestal, and he demonstrated to her how the union would work by grabbing a marking pen and pulling off its cap. Sally verbally signaled her comprehension ("I understand"). In addition, her eye attention, facial expressions, and physical copresence provided Mitch with assurances that she had heard what he wanted her to do, understood what he wanted her to do, and was willing to do what he wanted her to do.

Close proximity provided senior engineers with immediate and rather complete confidence that their junior colleagues were doing their work properly, safely, and in accordance with government regulations. It also ensured fluent communication among a colocated workforce whose members' proximity to one another facilitated planned and unplanned meetings, conversations, and visual awareness of who was in the office doing what. Close proximity afforded structural engineers ample opportunities for the teaching and learning that maintained the base of domain knowledge that the work required. To ensure these benefits of proximity, the structural engineering firms in our study maintained standard working hours, starting at 9 a.m. each day and ending at 5 or 6 p.m., a policy that facilitated synchronous interaction.

Hardware Engineers: Working Close Together to Coordinate

One might suspect that hardware engineering's highly digitized work environment—the consequence of their choice in the human-technology relationship to maximize the computer's role—would lend itself to the geographic distribution of work via advanced information and communication technologies. After all, software programming followed exactly that route, with many U.S. firms opting to send work to India, China, Canada, Israel, Ireland, the Philippines and elsewhere.[5] Certainly hardware engineering, with its modularization and textual coding, had come to look a lot like software programming.

The hardware engineers in our study, however, rejected this path, preferring instead to work in close proximity to one another. They neither telecommuted en masse (though one or two individuals worked from home) nor shipped work offshore to lower-cost engineers. Instead, they sat together in large rooms whose cubicle layouts placed them in easy walking distance of one another, and often within eyesight and earshot of one another as well.

As we have seen, the structural engineers in our study worked in close proximity so that senior engineers could, in the natural course of work, teach their junior colleagues and oversee their work. Senior engineers played no such role in hardware engineering, in part because liability concerns and government regulations were much less of a concern: The senior hardware engineers did not stand professionally or legally liable for their junior colleagues' work in the manner that the senior structural engineers did.

Moreover, because knowledge changed rapidly in hardware engineering, the senior engineers did not serve as conduits of fundamental knowledge to their junior colleagues. As a result, the senior hardware engineers were not revered by their juniors in the same way as the senior structural engineers were by theirs because having mastery over a large body of older knowledge was not useful in the context of everyday work. In fact, time itself seemed to take on a different dimension in the hardware engineering firms than in the structural engineering ones. Once, when referring to a file directory containing code for a previous version of Configurable Solutions's main microprocessor, Randy, a junior engineer, remarked, "Only the old-timers know how to run those files." The directory to which Randy referred was all of a year and a half old. In this environment of constant change, all hardware engineers, regardless of rank, had to constantly learn. Moreover, no hardware engineer was expected to ultimately possess the sum of the occupation's knowledge. Instead, each engineer, even the most junior, held some type of specialized knowledge; often, junior engineers, having just left the university, held some of the most up-to-date knowledge.

An electrical engineering professor at Stanford, an expert in design optimization and the creator of numerous software applications, explained to us why knowledge once viewed as fundamental in chip design was now not so:

For a long time, the old guys resisted the digital revolution [in chip design]. But it brought about a 10,000 × productivity improvement and put digital design in the hands of many, many more people. The old guard thought you had to know things like the size of the transistor, voltages, and currents, but now designers do not know fundamental concepts; they do not need to. These things are less important. For super-high-performance design—like the next Pentium memory cell—they still are important, but for most people working on most designs, they are not considered.

If hardware engineers did not require proximity for reasons of passing on fundamental knowledge or overseeing work, why then did they opt to work in close quarters? The answer is that the hardware engineers needed to coordinate with one another, and proximity facilitated that coordination. The comments of Salim, a senior manager at AppCore, about the high pressure on the engineers to get products to market quickly point to how the market and competition in this field shaped the choice to reject distant work by forcing a need for coordination in the form of collaboration and integration:

High-end design pushes the limits of performance. The fab technology required to make the chips, and sometimes the fab itself, does not even exist at the time we design the chips. The CAD tools needed to design them also often do not exist, and are built in parallel.... The stress level on these high-end products is very high.... Saturday is a mandatory workday. Right now, we are just finishing a product. The engineers were working around the clock and sleeping in sleeping bags in their offices. One engineer is a single mother; her two children slept in the boardroom one night. I found the children playing Monopoly there the next morning, and I told the woman I could afford a hotel room for them. For these groups, teamwork, energy, and conviction are instrumental.... If the product fails, it is for communication reasons, say a switch designed by one engineer fails to communicate with a switch designed by another engineer.... If we miss our deadline by six months, someone like [another firm] beats us to it. We depend on collaboration, integration, and teamwork, as well as our computer tools.

Our observational data support the professor's claim that fundamental knowledge had little enduring value in hardware engineering and the senior manager's argument that collaboration, integration, and teamwork stood at the heart of design in this field. Whereas the structural engineers in our study offered to teach a colleague just as often as they sought to learn from one, in the hardware engineering firms we studied the bulk of learning opportunities were sought, not offered. Moreover, the senior engineers were not the main teachers. In a reversal of the learning hierarchy that we found in structural engineering, the junior engineers were the most frequent teachers in hardware engineering and senior engineers were the least frequent. Notably, tutoring, in which the teacher adopted a pedantic tone and which occurred with some frequency in structural engineering, was largely absent in hardware engineering. In hardware engineering, the methods of choice when it came to teaching were providing information, collaborating, and instructing.

When providing information, the hardware engineers simply relayed facts, numbers, due dates, and similar information to the learner. Providing information usually happened quickly because the encounter required minimal instruction or guidance, as in the following example. Randy, an engineer at Configurable Solutions, and his officemate, Albert, were sitting in their shared office working on their computers. An engineer somewhere else in the company had reported a bug in Randy's code. Randy had just received an email about the bug via the firm's bug-tracking software.

Randy: [*Muttering as he reads the message*] Jim filed a bug on me! Every time he does, I can't understand what he's talking about.... [*Randy examines code that has been inserted into the email. He asks Albert about a variable in the code that he does not recognize.*] Albert, SysIOPAddr, what is it?

Albert: [*pauses on his laptop and turns his head toward Randy*] We have this concept of an I/O block now. The physical address is the start of the I/O block. All peripherals. [*Albert has just explained that SysIOPAddr, which stands for System Input-Output Physical Address, is a new variable that specifies the location of the input/output block.*]

Engineers provided information to each other in this manner all day long, adding to each other's knowledge in small increments. Proximity aided these exchanges, as the example of Randy and Albert makes clear: Randy had a question; Albert, just feet away, answered it for him immediately. In this case and in many others we observed, the engineer with a question did not need to do so much as lift a phone or type an email to get his answer. Typically, no more than a short walk to a colleague's cube was required for the engineer to receive a fast answer to his query.

Routinely, the hardware engineers we observed needed each other not for a fast answer to a short, pointed question but because one engineer had designed a component in the prior version of the microprocessor that another engineer had to update for the current version of the microprocessor. In other words, one engineer was the most recent past expert on the component; the other one was the emerging expert on that component. As we have noted, the hardware engineers began their designs by borrowing the code from the prior version of the microprocessor. Because the engineers rarely provided full comments in their code to explain what, how, or why the code operated, the engineers working on the newest version of the microprocessor had to call on the creators of the prior code to gain explanations. Although these conversations might have worked equally well over the phone, the fact that one engineer could stand and point at the code as it appeared on another engineer's screen was helpful in their discussions. Also helpful was being able to call on the other engineer on a moment's notice. The problems confronting engineers typically stymied their work; they might not have been able to make any progress with their code until they could speak to another engineer. In this manner, high product interdependence (the reliance of one product on some prior product) in this field shaped the engineers' choice to reject distant work. The fact that prior code was not self-explanatory also helped shape this choice. Being copresent provided the hardware engineers with the comfort of knowing that answers to their problems with prior code were only steps away.

In the example below, Luke at AppCore had called out over his cube wall to Ramesh for help. We met Luke and Ramesh in chapter 3, when Ramesh asked for Luke's help when analyzing clock paths. In the following conversation, which occurred two and a half months later, Luke was struggling with a gated clock path. Ramesh had written the code for this component in the previous version of the microprocessor, and now Luke needed to write it for the current version. However, the delay percentages in the wires had grown since Ramesh's design. As a result, the engineers needed precise estimates of clock path delays, which the constraint on time borrowing that Ramesh had imposed prevented. Luke had called Ramesh over to find out why Ramesh had made the constraint so restrictive. Ramesh now stood beside Luke, who sat in front of his monitor, examining the code in a text editor. The excerpt begins with Luke rendering his opinion of Ramesh's time borrowing constraint:

Luke: I think one nanosecond is a little hard on this.
Ramesh: This was for [the past version of the microprocessor]; I had fine-tuned it to one. This is not necessarily true for [the current version of the microprocessor] now because you have more gated clock paths. This is a different scenario now. [*They discuss why the current tools are not helpful for this problem before Ramesh continues.*] So what I had to do, I had to keep moving the setup time to some extent, where, sort of, do a binary search. Give it a high value, give it a low value, and then keep narrowing it down so that I get the most optimal. And that was one nanosecond for [the past version of the microprocessor]. I'm sure it is not the correct value here. This is much too hard for [the current version of the microprocessor], now. So you might want to reduce this one. And that's easy, right? The clock is getting set up, it is a parameter in the clock.dc file. [*He pauses.*] Because technically, it should be zero, remember, right? As long as the clock and the signal come at the same time, you are still okay. A little bit, maybe, earlier.
Luke: Yeah.
Ramesh: But there is no real setup required. [*Luke directs Ramesh's attention to his code, and they discuss preventing a spike before Ramesh continues.*] Okay, I'll tell you how you come to this value. The way you come to this value is by experimentation. And the experiment involves, you put this value as long as this is not the most critical path at the end of the synthesis run. Okay? So right now, you have overconstrained it so much that this ended up being the most critical path, and you were getting really a false violation at this point, right?
Luke: Yeah.
Ramesh: If you cut it down to say .5, and you see your most critical path is at .7 or something, then you know that....
Luke: Yeah, it is not really the critical path. That is why I want a memory path out of there, because the memory is at 1.35 or something like that.
Ramesh: I would suggest, well, this way of doing it, where it actually borrowed—
Luke: [*completing Ramesh's sentence*] —the right amount. It borrowed .9. It could have borrowed more and fixed that part of the path, but then this would have extra negative slack.
Ramesh: Yes, yes.

The conversation between Luke and Ramesh continued for another fifteen minutes or so, ending with Ramesh's advice to Luke for how to correctly set the time borrowing constraint. Luke needed Ramesh's assistance because Ramesh's old code did not explain why the constraint was set at its current level and under what conditions it would be enacted. The problem was too complicated for Luke to easily sort out by himself. Ramesh remembered how the constraint operated, and could provide Luke with guidance about how to modify the constraint.

On other occasions, hardware engineers needed each other because one engineer was the guru, or most knowledgeable person, about a technology in the design flow. Rather than trying to find the answer to their technology questions in whatever documentation (such as a manual) that may have come with the technology, the engineers preferred to pose their questions in person to the technology's guru. To limit interruptions of their work, technology gurus created FAQs that they posted on the firm's internal network to address the most common questions. Often, though, the engineers' questions were idiosyncratic, and, like as not, the problems prevented the engineers from continuing in their work. Consequently, they often desired an immediate, in-person answer from an expert on the premises.

An example arose one day when Fazio at Configurable Solutions approached Phil, the guru of a verification technology called 0-In, for exactly this kind of help. He had run 0-In, but it did not work properly. Phil asked him whether the clock gating was turned off. Fazio was unsure, so they began searching the code on Phil's computer. Looking over the code, Phil reported that the checker needed to be turned off, to which Fazio replied that as long as it was firing, he did not care, so he was just going to add a comment that it did not support gate logs or clocks. Phil told Fazio that his solution was no good, that he had to get rid of false firings, and that the way to do it was to change the comma to turn off the checker. Fazio returned to his cubicle and did what Phil had instructed him to do. This example shows that Fazio needed someone who was both knowledgeable about the software and capable of helping him debug his code, a combination that exceeded what a manual alone could provide. Turning to Phil was also no doubt quicker for him than turning to a manual would have been, though he interrupted Phil.

One day at a group lunch at Configurable Solutions, we asked the engineers about their email habits (they sat in cubicles within throwing distance of each other). Their answers reflected the kind of collaboration and real-time question and answers that our observations had revealed.

Researcher: You send an email, and if you don't get a response back within about thirty minutes, you ... [*pauses to prompt them to answer*].

Phil: You walk over....

Researcher: You walk over to the cube. [*The engineers laugh.*]

Phil: Yeah.

Researcher: You are fanatic email readers and you'll expect an email response rather quickly, so it's not really like you're willing to wait a day [for a response to an email].
Randy: And even if you did get it quickly, it still takes a half an hour of going back and forth to solve a simple problem sometimes. And when you walk over and talk to somebody, it only takes two minutes.
Brian: So then you're interrupting that person.
Phil: Yeah. It's a trade-off.
Fazio: We're interrupt-driven.
Collette: If it's something that I want an immediate answer, I don't send email. If I sent you email, then I'm comfortable [if] you send the reply to me later, not right away.
Fazio: —I mean, we only go see the person because we need him. And if we had to, the person's on vacation for a week or two, we just go into it, it takes us fifteen minutes more.
Raleigh: There are some things where … you just cannot do it.
Fazio: Yeah, if it is a very big thing.
Researcher: But someone goes on vacation for a week, it doesn't, [and] then you're stymied for a week? That really happens?
Raleigh: Sometimes.

Although the hardware engineers preferred to work in the office during the day to enable quick collaboration, rather than telecommute from home, they did employ advanced information and communication technologies to work from home after hours. As Salim from Appcore remarked,

> Many engineers have DSL lines at home. They work in the evenings, and sometimes all night.… The engineers have complete freedom over their time and how they spend it, provided the task is completed.

Most engineers with whom we talked did not work long hours in the evening and through the night as Salim suggested (or perhaps hoped), but they did routinely get up in the middle of the night to check on their simulations. By doing so, they could ensure that the results of finished tests would be waiting for them to analyze in the morning. Tests failed regularly for simple reasons, such as a typographical error or a syntax error in the command. These simple fixes were what engineers addressed in the wee hours of the morning. The engineers reserved more complex debugging for their normal hours in the office, in large part because that debugging was quite likely to require the aid of a colleague.

Although the hardware engineers wanted to work together, they did not work the regimented nine-to-five schedule that the structural engineers in our study did. Some hardware engineers who were the parents of small children came in at 7 a.m. so that they could leave in mid-afternoon to spend afterschool hours with their children. Most hardware engineers, though, arrived between 9 and 10 a.m. Few meetings were held before 10 or 11 a.m. for this reason; most meetings were held in the afternoon, as Eric explained to our observer in this excerpt from our field notes:

9 a.m. The office is very quiet and many of the lights are off. Eric says that before ten in the morning most people are emailing or reading "geek" newspapers. He opens up Netscape and shows me some of the geek newspapers.

Eric: [*navigating to www.tomshardware.com*] This is *Tom's Hardware*. A guy name Tom started it. It has PC-related news. It has news about the latest chips coming out. It is a news site, but not objective news. It is not like the [San Jose] *Mercury News*. It has more technical details. *The Register* also has technical news. [*He navigates to www.theregister.co.uk*.] It is not objective. Some of these websites have staff. Some of these websites have emerged out of a geek culture online.

Eric asks if I want to walk down the hall and find out if people are emailing and reading geek newspapers. I say sure. We walk down the hall directly in front of his office. We peer into each cube, but don't see anyone. At the end of the hall, we make a left turn, walk two or three feet, and then make a right turn. We walk up another hallway, heading back toward Eric's office. All the cubes are empty except one. The one man we see is emailing. We return to Eric's office.

Eric: There are a few meetings that start around 11 a.m. Actual joint collaboration does not begin until afternoon. Before noon, that is time for a coffee hit or to read email.

In short, what the hardware engineers wanted was not complete overlap in their hours but sufficient overlap (mostly in the afternoon and early evening) to allow coordination. Most important, they wanted proximity so that the answers to their questions were never more than a few steps away.

Automotive Engineers, Distributed Work, and the Siren Song of Labor Cost Savings

Automotive engineers were the only engineers in our study to employ distant work. For many years, IAC had acquired smaller automobile manufacturers in countries such as Australia, Brazil, Germany, Korea, Mexico, and Sweden to expand into those markets. At each of those sites, local engineers worked together to design and analyze vehicles for local markets. In short, the organization was global, but work on a single vehicle program was not. It was not until 2003, when IAC opened a captive offshore center in 2003 in Bangalore, India, to provide digital engineering services to its seven global centers, that IAC began to use advanced information and communication technologies to globally distribute engineering tasks within a vehicle program.

Unlike the other global centers, which were responsible for the design and analysis of specific vehicle programs, the India center did not "own," as engineers liked to say, any vehicle programs. When global center engineers needed help with their simulation models, they were to send work to the India center. Job requests were to indicate the type of modeling or analysis help needed and the job due date. Engineers at the India center would then complete the requested work and return the job to the global center from which it came.

Why did executives at IAC create an engineering center in India that would assist all the other global centers in their work but, unlike all those other centers, not have responsibility for the design, analysis, and manufacturing of any vehicles of its own? The simple answer was cost, particularly labor costs. Ken, a vice president for global engineering at IAC, explained:

> To stay competitive we have to keep costs down. One way to do that is to tap into big pools of engineering talent in places like India and China. All companies are making similar moves.

To be sure, the labor costs associated with an engineering center would be much lower in developing counties such as India than in the United States, Germany, or Australia. Our interviews and surveys with employees around the globe indicated that entry-level engineers in India were paid about one-fifth the average salary of engineers at the highest-paying centers and one-third the average salary of engineers at the next lowest-paying center in the company. Low wage rates did not, however, yield low-quality employees. Many of the engineers at the India center held master's and doctoral degrees from well-reputed Indian universities. Most of these engineers were trained in or for the aerospace industry, which pioneered the use of the types of simulation technologies that were commonly employed in automotive engineering. Despite a lack of experience in engineering cars and trucks, India center engineers were, as one manager mentioned offhandedly, "an exceptional value."

The labor cost rationale was also reflected in a phrase that we heard often throughout IAC's corridors: "twenty-four-hour engineering." The idea behind twenty-four-hour engineering was that, through offshoring, Indian engineers could continue, and perhaps complete, tasks overnight that their counterparts at the global centers had worked on during the day. As before, Ken was optimistic about the economic benefits of twenty-four-hour engineering:

When you have a truly global workforce you can do twenty-four-hour engineering. Right now we have a tremendous investment in IT and computing infrastructure that is being underutilized. Engineers in places like Bangalore can log into our servers while the workforce in the U.S. or Germany is gone for the evening, and continue to work. Tasks that were started during daytime on one continent can be completed in what was once "afterhours" during daytime on another continent. Not only will twenty-four-hour engineering help to reduce our costs, but it will help us to get products to market faster and with higher quality.

Ken's comment hints that although a labor cost rationale helps explain why IAC established the India center, the full answer is a bit more complicated. In a competitive industry in which profit margins on each vehicle sold were quite small, IAC's executives were keen to explore every option that might reduce product development time. Faster product development meant lower infrastructure and personnel costs per vehicle, auguring the capture of greater market share. Extensive studies that IAC conducted in the late 1990s suggested that engineering design and analysis work accounted for up to 70 percent of the time the firm spent on product development. Consequently, management highly prized a reduction in the amount of time engineers took to do design and analysis work, and focused their efforts on creating a leaner, tighter product development process. In this manner, factors of market and competition drove managers to employ offshoring.

Offshoring to India required a robust technological infrastructure. Not only did IAC need to ensure that its servers, databases, and network connections were robust, it also needed digitized work artifacts that engineers could send at a moment's notice through those electronic channels from the global centers to India. It was no accident, then, that IAC staffed its engineering center in India with analysis engineers, not design engineers.

As we described in chapter 3, design engineers required hands-on familiarity with and constant assessment of physical parts; managers needed to take but a single glance at design engineers' cubicles for a reminder of the importance of physical parts in design work. Although analysis engineers similarly worked with physical parts in the form of subassemblies and vehicles when validating their simulation models, in the eyes of IAC managers, analysis engineers worked almost exclusively with digital representations of parts. A glance around the cubicles of analysis engineers, as we recall from chapter 3, revealed no physical parts. That absence of parts reinforced managers' perception that analysis engineers worked almost entirely in a virtual world in which the engineers assembled virtual vehicles for virtual performance tests.

Hence, the IAC managers concluded that the digitized, computational nature of simulation model building and analysis meant that analysis engineers could easily share work across time and space by means of technologies that transported digital files. The comments of these three managers reflect that conclusion:

—It wasn't like you were going to take a bumper and send it to a different country and say, "Here, run some tests on this." That just wasn't feasible. Bumpers are expensive to ship and the facilities that you'd have to build to test them are even more expensive. But now that we work in math, you can do that. It's like sending a bumper, but the postage is free.

—When you look at our portfolio of engineering expertise, it makes the most sense to offshore [analysis] engineering work because, unlike design engineers, [analysis engineers] will—at least someday, we hope—work in a completely virtual world to do their model building and analysis. So it just makes sense that we get them the technologies they need, the information systems to coordinate their work, and then they'll be able to divide up their work and send it around the world. That's the beauty of using these technologies.

— When we were thinking about how to scale our simulation work, we realized that we had high fidelity in our models. So you didn't need to go to the proving grounds each day to map what you were seeing on the screen onto reality. We looked around and said, "we've got lots of bandwidth with our Internet backbone." So it makes sense that we can take things that we could only do in Michigan just a couple years ago and do them anywhere in the world. Maybe even in places where our labor costs weren't quite so steep. It seemed obvious to go to India because all the information anyone needed to do the work was in the model.

The decision to locate the center in Bangalore was telling. With its large information technology infrastructure, Bangalore was the offshoring, though not the automotive, capital of India.[6] The choice of Bangalore and the restriction of work in India to purely math-based services, with no physical testing involved, reflected management's vision of the India center as a step toward its goal to replace more physical tests with virtual tests.

When engineers at the global centers learned of management's plan to create a captive offshore center for analysis work, they were at the same time excited and skeptical. They were excited, as Adam, an engineer in Australia, remarked, because this plan meant they could send the most boring and tedious parts of their jobs overseas:

When they initially told us we would have the India center as a resource everyone was saying, "This is so great." In a lot of professions people are worried that their jobs are going overseas. We know that is not going to happen here because the whole design and analysis of vehicles is so complex that you can't just have them do it some other country. And India was not getting [vehicle] programs either. Even if they did someday, they would be Indian-centric programs, which would result in vehicles that would not sell well in places like Australia or Germany, so they would always need centers here. What it did mean was that we could decide what work to send to India. And people loved this idea because, well, as they say, "Shit flows downstream." I mean, you can send the stuff you have no interest in doing to India—the boring stuff—and then when it comes back you integrate it with your model. That is something we were all very excited about doing and happy to have India as a resource for.

Underlying this excitement, however, was a guarded skepticism. Analysis engineers at the global centers knew that they relied tremendously on observing physical tests and inspecting parts after those tests to make sure that the simulation results correlated strongly with the results of physical tests. They worried that the absence of infrastructure to house physical parts or to physically test vehicles at the India center would prevent engineers there from similarly validating their models. The result, they feared, would be faulty models. By constructing no physical testing facilities at the India center, the IAC managers seemed to have come to trust simulations far more than did the engineers who produced them. In other words, managers appeared willing to cast off the logic of belief discussed in chapter 3 to follow the siren song of labor cost savings proffered by the offshoring of analysis work. In accordance with management's wishes, analysis engineers formally divided their work process into nine sequential tasks, which box 5.1 displays.

Box 5.1

Nine Tasks in Building and Analyzing a Vehicle Simulation Model

1. Build a mesh from design engineers' master CAD files of parts.
2. Set up the model to be run with the mesh with boundary conditions and accelerometer placements.
3. Run the simulation model.
4. Analyze and interpret the simulation results.
5. Correlate these results with the physical test results to validate the simulation; modify the model if needed and repeat tasks 2 through 5.
6. Develop improvement ideas for design based on validated simulation results.
7. Create case studies (additional simulations) to test the improvement ideas.
8. Analyze and interpret the results of the case studies.
9. Make recommendations to change the design of parts (and, ultimately, the master CAD files).

Analysis engineers reasoned that tasks 1 through 3 (building a mesh, and then setting up and running a simulation model) could be sent to India, but subsequent tasks (analyzing and validating the results, and then making recommendations) should remain at the global centers. Analysis engineers viewed building a mesh as the least interesting part of their work; they also thought that access to physical parts might not be needed until the later steps of analyzing and validating a simulation. As Gretchen, an analysis engineer who in chapter 2 told us how building a simulation model was a mix of art and science, explained:

> We do lots of tasks. Most of what we do is analysis, but a lot of our time is just model building. That's routine, sort of standard stuff. It's also not so detailed. What I mean is, you're working to build a mesh and you're deciding on how to shape and refine elements and connections. That's all abstract stuff. So, you don't need to look at parts or see tests because real objects don't have elements—they're not divided into boxes for computation. It's just more removed, at that point, from the actual vehicle.

Once the India center was established and offshoring commenced, analysis engineers routinely sent the first three tasks to India, but almost never sent later steps in the simulation process. What this meant in practice was that the engineer in India did most of the grunt work to convert (through meshing) the design engineers' CAD files into a simulation model, to set the boundary conditions to the model appropriate for the particular analysis to be conducted with it, and finally to run a preliminary test of the model to confirm that the model was set up properly. This last step—making sure the model was set up properly—was often quite difficult to do. To ensure that a model was set up properly, an engineer would submit it to the solver—the code that computed and solved the equations embedded in the model—for a test run to see whether the solver would run the model or kick it back to the engineer with an error message saying that the model had internal problems.

One reason (of several possible reasons) that a model could fail was that it contained too many "penetrations," a situation in which one part jutted through another part. Impossible in the physical world, penetrations littered the virtual world. Penetrations occurred because design engineers frequently altered their part's design, which then required changing their CAD files and redefining the spatial coordinates that specified where that part was located in the vehicle. Changes in the size or shape of one part typically had ramifications for adjacent parts. But because design engineers often neglected to tell each other about changes, the engineers responsible for adjacent parts failed to alter their designs, resulting in a penetration.

Analysis engineers needed to resolve penetrations before running their models because a vehicle with components jutting through each other behaved differently in a crash, for example, than one whose components did not exceed their allotted space. Resolving all penetrations was infeasible: penetrations were so plentiful that their complete eradication would have caused the analysis engineer to miss his program deadlines. Thus, an analysis engineer had to decide which penetrations to correct. Most engineers opted to correct only those penetrations that involved parts that were implicated in the test they intended to run and ignored those penetrations that would not meaningfully affect the simulation. Analysis engineers knew which parts of the vehicle were implicated from observing tear-downs after physical tests. Gordy, an engineer from the U.S. center, explained how he decided which penetrations to correct:

> Well, you just know from seeing the [physical] test. You go see the vehicle after the test and you look and see what areas of the vehicle were impacted—you see what was damaged and what wasn't. You can even touch the parts to get a good sense of it. That gives you a sense of the basic area of the vehicle that will be your load path. That's the area that you have to focus on. All the other areas you can leave because they're not so involved. You just look at test after test after test, and you eventually learn what your model should be like.

Although the Indian engineers were technically competent, when fixing penetrations they encountered problems for at least three reasons. First, because automobiles were not nearly as common in India as they were in the countries of the other global centers, the India center engineers lacked the cultural familiarity with automobiles present at the other centers. Most of the India center engineers had never driven, much less owned, an automobile. The Indian engineers realized that their unfamiliarity with automobiles hampered their ability to deal with penetrations because they did not know how each part was supposed to look.

The second reason why the India center engineers had difficulty when trying to fix penetrations was that the India center had no physical testing facilities; consequently, its engineers could not inspect physical parts or the results of physical tests to determine which penetrations in virtual parts they should fix. Faced with such uncertainty, the engineers sometimes resorted to guessing which parts might be implicated in a particular analysis. In other instances, Indian engineers chose not to fix penetrations. In general, because the choice of which penetrations to fix and which to leave was unguided by an examination of physical parts and the results of physical tests, Indian modelers often left penetrations that would later derail analysis.

On other occasions, Indian engineers chose to fix all penetrations, which led to the third reason for their problems: in taking this approach, they became vulnerable to the coordination and communication difficulties that students of geographically distributed teams repeatedly describe.[7] To build a simulation model, Indian engineers needed the design engineers' master CAD files. But whereas the engineers at global centers had access to DataExchange, the Indian engineers had access only to the files that the engineers from the global centers placed on the FTP site for them. Getting the most recent files required the Indian engineer to ask the sending engineer to locate and upload the files. Furthermore, determining which part to alter when correcting penetrations required negotiation between the design engineers responsible for the parts. Because the Indian engineers had no direct contact with the design engineers, they resorted to the guessing described above or routed their queries through the sending engineer. Either strategy could and often did lead to delays and mistakes.

Although problems such as those that arose in the context of penetrations proved common in the offshoring of analysis work, IAC management had no intention of shuttering the India center. Instead, management expected engineers to figure out how to make the situation work, with most of the burden falling on the India center engineers. These engineers knew that they needed to come up to speed, and the engineers at the global centers recognized that their Indian colleagues lacked the skills and knowledge the job required. Engineers at the global centers did not have the time, however, to walk India center engineers through each step of the model building that needed to be done, nor did they have the time and inclination to take pictures of parts or crashed vehicles to send to India. As Fernando, an engineer at the Mexican center, said:

> If I were going to spell out how to do everything, it would just be faster to do it myself. And I can't always be going to the proving grounds for things they [the India center engineers] ask about. They need to use the models we send them as guides and learn from those. Those models are based on what we saw at proving grounds in previous versions of the product. Hopefully those models are helpful to them to learn.

As Fernando indicated, the most common (and often the only) way that engineers at the global centers helped the Indian engineers come up to speed was to send them prior simulation models with the hope that Indian engineers would be able to discern the steps and decisions implicit in the older models and follow them when building and analyzing the requested, new models. Analysis engineers at the global centers saw knowledge transfer via these "reference models," as the prior models were called, as advantageous because reference models helped to overcome problems of communication with engineers who lacked knowledge and who worked at a distance. An analysis engineer named Gary at the U.S. center observed,

> We almost always send reference models with the jobs we package for India because they help show them [the India center engineers] how to build a model a certain way or do a certain kind of analysis. They can just look at the reference model and examine it to see what we did on past products. We have to send them because the Indian engineers are new and they don't know how to do this kind of work yet. We can't really teach them because they're so far away and in a different time zone, and they're at work when we're sleeping. So the reference model gives them something to copy. If they copy them on their tasks we send them now, they'll get the knowledge they need to do the work, and we don't need to write some lengthy document about how to do it.

By sending reference models to India, analysis engineers at the global centers hoped that engineers at the new India center would quickly gain the knowledge they needed to work on current vehicles through replication. Although the Indian engineers found these reference models helpful, the reference model sent to India from a global center was useful for explaining what needed to be done only if the Indian engineer using it could understand why the engineer who created the reference model had taken the steps he or she did. The following example outlines how this difference between knowing *what* and knowing *why* became so crucial.

Dipal, an analysis engineer at the India center, was assigned a job for the seventh generation of one of IAC's midsize luxury vehicles. With each generation of the vehicle, the architecture had changed moderately. Moderate changes in a vehicle's architecture could easily cascade into major changes in its structural elements, which was what had occurred between the sixth and seventh generations of this vehicle. Arnold, the U.S. engineer who assigned this job to Dipal, had sent brief instructions on the model building and preliminary analysis he wanted Dipal to do. Arnold also sent a reference model based on a simulation from the sixth generation of the vehicle. Figure 5.1, which illustrates that reference model, shows a vehicle with varying mesh sizes (e.g., a fine mesh on the front of the vehicle, a course mesh on the rear).

Engineers varied the mesh sizes so that they could achieve fine-grained results for the vehicle areas of most interest to them, given that limited computer processing power did not permit fine-grained results for the entire vehicle. Arnold hoped that this prior model would allow Dipal to under-

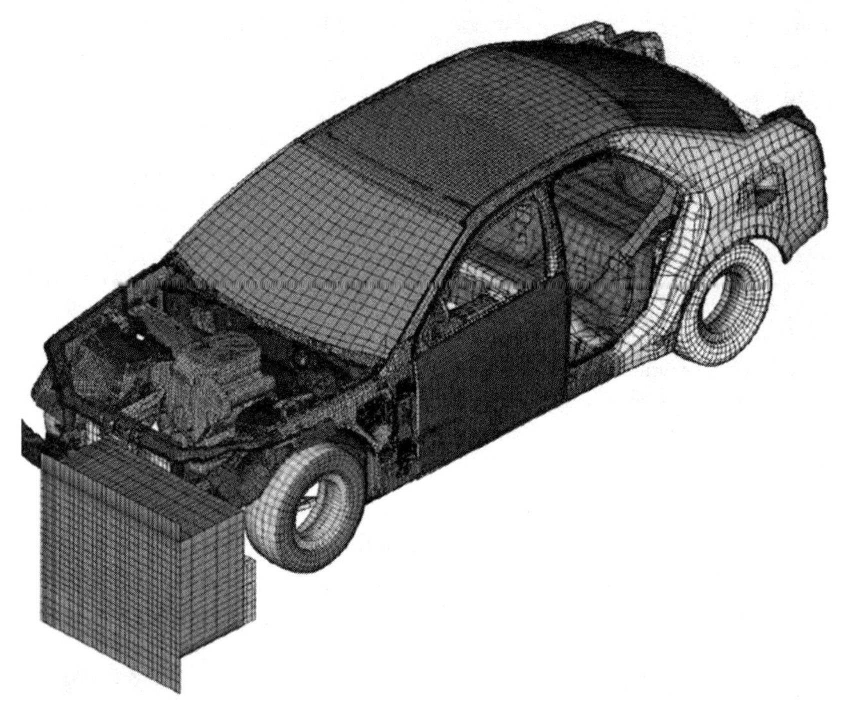

Figure 5.1
Reference Model for Offset Deformable Barrier Crash Simulation Analysis.

stand how to build the model for the current generation of the vehicle, as he explained to our observer:

This reference model [*he points to the screen*] has everything he [Dipal] needs to know to build the simulation model. When he looks at it he'll see which areas of the model are most sensitive for the ODB[8] analysis, and he can adjust his new model [for the seventh generation] so the appropriate areas have a finer mesh. I like sending these reference models because it's way quicker and easier to convey the information. If I didn't have it I'd have to write out instructions, like "mesh the cradle at 5 mils on the driver side and 8 mils on the passenger side," and that would take a long time. Or I'd have to call him up and tell him, which would be hard to 'cause he's in India and he'd have to like stay late or something to get the phone call, and that still wouldn't be that good because we're not next to each other so I can't point to parts on the model. That's why the reference model is better.

When Dipal reviewed the reference model that Arnold had sent, he determined that he could not copy it exactly because the geometry of the seventh-generation vehicle was substantially different from that of the sixth-generation vehicle. In other words, he concluded that he could not vary the mesh densities in exactly the same pattern that Arnold had. Dipal reasoned that he needed to understand the principles behind what Arnold had done so that he could make decisions about how to vary the mesh densities for the new generation of the vehicle. He explained his concerns to our observer:

> I see that they used a finer mesh for the lower trim pieces than for the upper trim pieces. But in this model [from Arnold] there are different structural connectors between the uppers and lowers than in the [seventh-generation] vehicle model. So I am not knowing if the loads will be different through the body-in-white,[9] and I am not sure if I should be meshing them the same or differently.

Dipal was saying that he did not know whether the seventh-generation design would dissipate crash energies differently from the way the sixth-generation design did. He did not have sufficient product knowledge to make a decision. As Dipal commented,

> Looking at this reference model was very useful because until I did it I did not know what I did not know. I mean, from looking at it I realize now that I do not know why they would mesh these uppers and lowers at different sizes, and that shows I do not know how the load moves through the vehicle in this [the seventh-generation] design. So I must ask the engineer so I can learn why they did it this way, and then I can come to know how to do it for the present model.

The next day, Dipal and Arnold talked by phone. Dipal told Arnold that, after reviewing the reference model, he realized he did not fully understand why the mesh sizes varied the way they did. Dipal said that because he did not know how crash energies would move in the seventh generation vehicle, he had decided that his safest bet would be to finely mesh both the upper and lower trim pieces. Arnold replied that doing so would too greatly increase processing time, and explained what kinds of geometric changes would warrant a fine mesh. Reflecting later on this interaction, Arnold commented to the observer, "I didn't know he didn't get it about the energies in the model. I'm glad he called me."

As this example makes clear, reference models helped engineers at the India center realize what they did not know. Once they came to realize what they did not know, they could seek out engineers at the other global centers to learn the principles and decisions implicit in the reference model. The India center engineers needed this information to knowledgeably use the reference models as guides for their models. But to do so, Indian engineers had to work hard to seek out their counterparts at the global centers and learn from them. Even if the Indian engineers had the initiative and foresight to do so, they had to hope that engineers at the global centers who sent them work were available and willing to teach them. In other words, teaching and learning across distance began to characterize analysis engineering work at IAC, a consequence of offshoring that no one predicted. Unfortunately, not all engineers at the global centers were keen to teach their Indian center counterparts.[10]

At the root of the problem of offshoring was managers' assumption that analysis work was completely virtual, and as such did not require for its accomplishment physical parts or physical testing facilities. That advanced information and communication technologies enabled automotive engineers to send digital artifacts halfway around the world for their completion did not mean necessarily that the engineers should have done so.

Occupational Approaches to Locating Engineering Work

We learn from the accounts in this chapter that occupational factors strongly shaped how the engineering managers we studied determined where to locate their engineering workforce. The managers in two occupations made a similar choice—to keep engineers proximate to one another—but they did so for different reasons. The managers in the third occupation made a different choice, as reflected in the distribution of engineering tasks to engineers around the globe. Although advanced information and communication technologies facilitated the choice to disperse work in this third occupation, they did not mold that choice. Instead, across all three occupations, engineering managers made choices based on factors that were unique to their occupation, despite similarities in the underlying work activities of their engineers, the similar affordances of their technologies, and the complexity of their products.

In structural engineering, managers chose to locate engineers in close proximity to one another largely because, in their field, knowledge was enduring and vast. Each engineer was expected to know all there was to know, a feat that required years of training and experience in the field after one's university education. Liability concerns and government regulations were paramount and were related to the educational imperative, in that licensing requirements institutionalized senior engineers' role as teachers of their junior colleagues, who were deemed ineligible to complete designs on their own until they had spent years learning with a master engineer. Structural engineering's choice to limit the role of the computer in this field further cemented the need for proximity: large physical drawings that served as the centerpieces of design discussions and daily work reviews would have been difficult to transport over distances and, when viewed on the computer screen, hardly would have enabled the kind of group discussion and collaborative thinking that senior engineers sought for their project teams. In sum, the structural engineers worked in close quarters because junior engineers needed senior engineers to show them how to do the job.

In hardware engineering, pedagogical concerns did not motivate managers to locate engineers in close proximity to one another. In this field, knowledge was fleeting, not enduring, a fact that significantly reduced the value of being senior. The amount of knowledge required to produce a chip was vast, but no single engineer was, in his lifetime, expected to hold all of it. Rather, engineers divided up knowledge areas such that each one became a guru in some topic, be it a technology or a chip component or a process. Nor did liability concerns and government regulations drive the hardware engineering managers' choice to colocate engineers; these engineers did not hold professional licenses, and anyone could design and sell a chip if they so desired.

What drove the engineering managers to place their engineers in close proximity to one another was product interdependence. Because the engineers began their designs with code from the previous microprocessor, they needed to understand how that code worked. In the absence of good documentation of code, the engineers had little choice but to turn to each other for explanation, which we observed them do time and again. Moreover, perhaps because they could not resist a good puzzle, perhaps because they knew that doing a good deed one day would earn them a good deed in return the next day, or perhaps out of compassion, hardware engineers routinely helped each other solve tough problems, staring at a colleague's screen and patiently working through with him the ins and outs of the code until a solution, a theory, or a good temporary patch revealed itself. In sum, the hardware engineers worked in close quarters because engineers of all ranks needed one another equally to get their work done.

In automotive engineering, the managers chose to distribute work around the globe. Their desire to do so was fueled by the siren song of labor cost savings and the potential of reducing product development cycles. A shorter product development cycle meant they could get to market earlier; coupled with lower costs, an earlier market entry would help them capture greater market share and fare better in a highly competitive industry. Their ability to lower costs and shorten product development cycles was enabled by an increasingly virtual world of work in which mathematically based simulation models might someday substitute for expensive and time-consuming physical tests. Although the managers believed that simulation was ready to replace physical testing, the automotive engineers were more skeptical because they knew just how often they had to validate the results of their simulations against physical test results.

To comply with upper management's mandate to send work to India, analysis engineers identified early tasks in the process that they believed might not require access to physical parts and physical tests, and later tasks that did. However, analysis engineers underestimated the extent to which even the most basic and mundane model building tasks required access to physical parts. As a result, the India center engineers had to maintain regular contact with the engineers at the global centers not only to gain physically derived knowledge from them but also to learn why engineers had made the modeling decisions they did. In sum, global distribution carried important consequences for how engineers needed to communicate with other and how they needed to teach and learn from one another to make global product development work.

Each of these three accounts of technology choices for where to locate engineering work speaks to the influence of occupational factors in shaping those choices. The strong influence of these factors on the choices that engineers made with respect to the human-human relationship, like the choices that engineers made concerning the human-technology and technology-technology relationships discussed in chapters 3 and 4, respectively, is problematic for existing explanations of technology choices and points to the potential value of an occupational perspective.

6 The Role of Occupational Factors in Shaping Technology Choices

Our thesis in this book is that occupational factors strongly shape technology choices in the workplace. In this chapter, we bring together from prior chapters the threads of the occupational factors that arose in our analysis of technology choices in the three engineering occupations that we studied. These factors included product liability concerns, the rate of change of occupational domain knowledge, market structures, the ability to reduce product complexity, the division of labor, and other factors discussed in chapters 3, 4, and 5. In this chapter we summarize which factors affected which choices in which engineering occupation. The overall picture that emerges is one in which various occupational factors were differentially at play in each occupation. We show that the factors that influenced technology choices differed by occupation, not by individual or organization. Further, we show that, within an occupation, the factors at play differed by the type of relationship (human-technology, technology-technology, human-human). We tentatively explore the predictive potential of each factor by examining its influence across the three occupations that we studied.

How Occupational Factors Manifested across Engineering Occupations

Table 6.1 delineates the manifestation of factors by occupation. For example, it depicts whether an occupation had a slow (structural engineering), moderate (automotive engineering), or fast (hardware engineering) rate of knowledge change. The first three rows of table 6.1 remind us that three occupational factors—type of work, type of technology, and product complexity—were constant across our sample. As we explained in chapter 2, all the engineers we studied, whether they were structural engineers, hardware engineers, or automotive engineers, carried out the *work* of product design through design and analysis activities, such as sketching, calculating, modeling, and testing. Similarly, all of these engineers employed advanced information and communication *technology* in the form of sophisticated computational, logic, and graphical computer applications to aid them in their design and analysis activities. In addition, *product complexity* was high in automobiles, buildings, and chips, with each product consisting of tens of thousands of components whose smoothly integrated functioning was essential for satisfactory product performance. That the three occupations in our sample were alike in these three factors permitted the comparisons that we made in our study by providing a common base, allowing us to ask, "If the type of work, the type of technology, and product complexity were identical across these occupations, why did the engineers' technology choices differ?"

The answer to that question lies in the remaining twelve occupational factors that complete table 6.1. This table summarizes the occupational factors that we introduced in chapters 3, 4, and 5, factors that arose naturally in our discussions of technology choices related to the human-technology, technology-technology, and human-human relationships, respectively. The twelve factors that appear in the remaining rows in table 6.1 either emerged from our data (for example, our informants mentioned them directly) or arose in our analysis (that is, we inferred from our observations of engineers at work the influence of factors, which we then confirmed by interrogating our data more completely or interviewing engineers and their managers).

In chapters 3, 4, and 5, we provided quotations and excerpts from our field notes as evidence of this grounding of the factors in our data when discussing choices in the realm of each relationship. Scholars call this approach "grounded theory"[1] because it begins with data and builds toward theory based on what the data yield, rather than beginning with theory and evaluating data in light of the theory. Thus, in employing the approach of grounded theory for our analysis of the factors that shaped technology choices, we did not begin with a set of a priori factors deduced from the literature, which we then set out to investigate. Instead, we looked for factors in our data. Thus, if some factor appears to be missing from this set, it is because that factor did not arise as a possibility either during the course of observation and interviewing or in our later analyses. We provide more details of our use of grounded theory in the appendix.

We classified and organized the twelve occupational factors into five categories: environment, knowledge, work organization, product, and technology. We accomplished this classification by asking what the focus of the factor was, or what lay at its core. For example, when we talk about the "self-explanation of work artifacts," we mean the extent to which artifacts revealed the knowledge embedded in them; hence, knowledge lies at the

Table 6.1
How Occupational Factors Manifested, by Occupation

			Structural Engineering	Hardware Engineering	Automotive Engineering
Occupational Factors	Common	*Type of work*	Design and analysis	Design and analysis	Design and analysis
		Type of technology	Sophisticated computational, logic, and graphical computer applications	Sophisticated computational, logic, and graphical computer applications	Sophisticated computational, logic, and graphical computer applications
		Product complexity	High	High	High
	Environment	Market	Custom production	Mass production	Mass production
		Competition	Bid-based	Time-based	Time-based
		Liability and government regulations	High	Low	High
	Knowledge	Rate of knowledge change	Slow	Fast	Moderate
		Testing capability	Minimal	Considerable	Considerable
		Self-Explanation of work artifacts	Low	Low	Low
	Work Org.	Division of labor	Simple	Simple	Complex
		Task interdependence	Low	Low	High
	Product	Product interdependence	Low	High	High
		Complexity reduction ability	Considerable	Considerable	Minimal
	Technology	Technology cost	Low	High	High
		Technology transparency	Low	High	Moderate

core of the factor, and we classify it thus. This classification is useful because it helps us to think not only in terms of individual factors and how they shaped choices but also in terms of how sets of related factors may have done so, possibly in combination. In our descriptions of the factors below, we draw on material we presented in chapter 2 when detailing the work of each occupation and in chapters 3, 4, and 5 when discussing technology choices and what shaped them. Thus, nothing in these descriptions should strike the reader as new, but this chapter is the first place where we have turned the focus of attention directly on the occupational factors and compared their influence across the three occupations.

Environment Factors. Three factors characterized the environment in which the occupations were situated. The *market* factor refers to how the engineers designed products for customers. The structural engineers designed idiosyncratic products for individual customers. Each building was a custom design, even though many aspects of the design may have been typical within a certain type of structure (such as commercial real estate or semiconductor manufacturing facilities). Each product in this field came with its own set of drawings, its own design, and its own analysis, all prepared for a single client. The opposite was true in hardware engineering and automotive engineering, where engineers created products for a mass market. In these fields, the engineers created a design and performed analyses for a product that they expected would be replicated many times over for customers who would see little variation in the final product.

The *competition* factor reflects what the basis of competition was among the firms in which the members of these occupations were employed. In structural engineering, firms competed through bidding to win projects for individual clients. Once the bid was awarded, the engineers worked to meet deadlines established in the bid, but they did not work with the constant thought that they were competing against engineers in other firms to complete their design first. That situation, however, was exactly the one faced by the hardware engineers, who competed on the basis of time (as measured by the calendar) against other firms to arrive first to market with a product. Because the time of market arrival was a good predictor of market share, the hardware engineers lived with the everyday worry of not being fast enough. Automotive engineers worked toward deadlines set internally in conjunction with normative standards of product release (the annual fall season of new models). Although this timing was standard across the industry, the automotive engineers nonetheless competed on the basis of time. Not only did they want to beat their competitors to market in any given year, they also wanted to reduce their product design cycle, which spanned three to five years, so that they could shave one or more years off their time to market. Thus, like hardware engineers, automotive engineers competed on the basis of time.

The third environment factor across the three occupations is *liability and government regulations*. Structural engineering and automotive engineering were marked by high product liability as well as by government regulations often aimed at consumer safety protection. If buildings fell down or automobiles broke down (or, worse yet, performed poorly in crashes), their owners were likely to suffer injury and to sue the makers of the product. In the case of structural engineering, product liability and government regulations were explicit in the licensing requirements of engineers. These requirements specified tests and qualifications for practicing engineers; they permitted only licensed engineers to stamp final design drawings to indicate approval of, and responsibility for, the design. State building codes and county permit reviews provided further evidence of this factor. In automotive engineering, no such professional licensing requirements existed, but rafts of lawsuits and consumer groups made abundantly clear to engineers the product liability issues at stake. Moreover, automotive firms were prevented from selling any products that failed to meet stringent government safety and other standards, with the government conducting its own tests to ensure product compliance. Hardware engineers, by contrast, were little affected by product liability concerns, perhaps because product warranties were typically sufficient to cover the consequences of product failure. Similarly, they faced no substantial government regulations concerning the safety of their products. Hence, liability was low in hardware engineering.

Knowledge Factors. The first knowledge factor is *rate of knowledge change*. In structural engineering, knowledge is enduring. For example, the load that a beam 18 inches deep and 35 pounds per linear foot could bear is the same today as it was a decade ago. Similarly, much of the knowledge of how buildings stand up and why they fall down is based on Newtonian principles, which have been known for nearly three hundred years. True, new materials, new construction methods, and new technologies regularly appear, but the bulk of fundamental domain knowledge in this field is static and the rate of change is slow. In hardware engineering the opposite is true: domain knowledge changes rapidly, and the value of any given bit of knowledge decreases considerably with time. Not only do processes, materials, and technologies rapidly evolve, so do the fundamentals of what designers need to know. In this occupation, the rate of knowledge change is fast. In automotive engineering, the rate of knowledge change is moderate, falling in between the extremes of structural and hardware engineering.

Although materials, manufacturing processes, and vehicle features are continually evolving, such that the vehicles of the current period are distinct from those of a decade before, the state of the art is likely to remain stable over the course of the several years that span a vehicle program's development cycle, such that engineers are not constantly struggling to keep up with change in their domain knowledge. Moreover, despite a strong push toward computer simulation, the continued value of physical parts and physical testing contributes to the enduring value of experience, such that long-tenured engineers possess useful and respected knowledge.

The second knowledge factor is *testing capability*, or the extent to which engineers in an occupation were able to determine the soundness of their designs. This capability was considerable in automotive engineering because the engineers could build physical models of their vehicles and test them in a variety of conditions, including ones that presumably closely mirrored actual use. They drove vehicles into walls, slammed vehicle doors thousands of times, and took vehicles out on overnight road trips. Hardware engineers similarly enjoyed considerable testing capability because their designs were implemented in binary language of ones and zeros, which afforded clear tests of failure. Simply put, code either worked or it did not; any failure that engineers could imagine, they could devise tests for, and for the remainder, they had a strong confidence that random testing would uncover any problems. Structural engineers stood alone in our sample in having minimal testing capability. Whereas automotive engineers could run prototype vehicles into walls and hardware engineers could write a battery of tests to run against their component code, structural engineers were not in the habit of constructing buildings and then subjecting them to harsh conditions of earthquake, wind, blizzard, or other natural events. Moreover, their models rested on assumptions that could not easily be tested because equally knowledgeable engineers may have debated, for example, which way the load would travel through a section of a building. Structural engineers lacked the means of clear physical confirmation (as with the automotive engineers) or clear virtual confirmation (as with the hardware engineers) of the soundness of their designs.

The third knowledge factor concerns the *self-explanation of work artifacts*. By this phrase, we mean the extent to which engineers' everyday work products of design and analysis—sketches, calculation sheets, models, simulations, and the like—make evident to individuals beyond their creators the implicit knowledge residing within the artifacts. Self-explanation of work artifacts was low in automotive engineering. As an example, when constructing a simulation model of a vehicle, an automotive engineer made any number of decisions related to which parameters to choose for variables and options within the modeling software interface; the resulting model reflected the choices made, but not the reasons for those choices. No mechanism for recording decision rationales existed within the technologies, and even if it did, it seems unlikely that busy engineers would have taken advantage of that feature. Hardware engineers similarly could have easily added comments to their code if they so desired, but few ever did. Consequently, their code files reflected the same low levels of self-explanation as the automotive engineers' simulation models. Structural engineers shared this fate as well: their final designs and calculations displayed the components that they chose and the performance of those components, but not the reason for the selection of those components. In each occupation, engineers who examined the work artifacts of other engineers had difficulty recovering the implicit engineering knowledge of the artifact's creator that the artifacts embodied.[2]

Work Organization Factors. The *division of labor*, the first of two work organization factors, was fairly simple in structural engineering and hardware engineering. In these fields, each engineer engaged in both design and analysis; additionally, managers grouped engineers into project teams in which engineers had similar tasks and similar technologies but were responsible for different components. For example, structural engineers in a group might be assigned the design of different floors of the same building and hardware engineers might be assigned different components of the same microprocessor. In automotive engineering, the division of labor was much more complex. Here, managers separated design from analysis, and within each of these fields further divided engineers into smaller and smaller groups. As a result, engineers working on the same product had different tasks and used different technologies.

Structural engineering and hardware engineering were also similar in terms of having low *task interdependence*, the extent to which one engineer relied on one or more others to complete his or her work. Modularization of the product (into floors, into microprocessor components) with assumptions of, or specifications for, how all the parts would operate in the whole allowed engineers in these occupations to proceed with their work on their components without having to wait for the completed design and analysis of other components completed by other engineers. In automotive engineering, by contrast, task interdependence was high. Although it was true that design engineers worked on their parts largely independently of their peers who were working on their own parts, an analysis engineer who wanted to assemble a mesh of the vehicle for simulation purposes could not proceed without designs in the form of CAD files from the design engineers whose work preceded his. Similarly, a second analysis engineer who wished to employ the first analysis engineer's mesh in a simulation model could not do so until the mesh was complete. In analysis engineering in particular, an engineer's ability to start work was governed by the work of others, a situation that did not hold true in the other two occupations.

Product Factors. Although task interdependence was low in hardware engineering, *product interdependence* was not. By product interdependence, we mean the extent to which the creation of one product (not component) is dependent on the creation of some other product. For hardware engineers, product interdependence was high because almost all microprocessors were new versions of some prior version of the same microprocessor. Thus, hardware engineers did not begin their work on a new microprocessor with blank screens and empty files; rather, they began by downloading the prior code files associated with the prior microprocessor, and they set about modifying that code. Thus, each new product was highly dependent on the version that came before it, and, by extension, on the entire string of products in that line over time.

The situation was similar in automotive engineering, where the majority of vehicles were new generations of previous models of the same vehicle. As a result, automotive engineers began their work with the CAD files of previous vehicle models and modified those files based on new product specifications. Analysis engineers looked to simulation models of the prior generation of their vehicle to understand how they should set up the model for the current generation. Consequently, for automotive engineers, product interdependence was also high.

Structural engineers were unique in our study in that, for them, product interdependence was low. We recall here that engineers created custom products for individual clients. Although many parts of a building resembled other buildings, thereby allowing structural engineers to borrow some prior work for current projects, a good portion of each building design was idiosyncratic to the current project and required a custom solution. For this reason, the structural engineers, unlike their hardware engineering and automotive engineering counterparts, typically began work on a new building with a blank page and a blank screen.

A second product factor is *complexity reduction ability*. By complexity reduction ability we mean the extent to which engineers can cope with, or work around, the complexity of their product. The structural engineers' ability to reduce product complexity was considerable because they could make simplifying assumptions, as when they drew a 45-degree line across a rectangular floor to assume the design on one side of the line would largely mirror that on the other side, or when they assumed that the third floor in a skyscraper would experience loads in a manner similar to the fourth floor. They also employed approximate analytical methods and 2-D representations to reduce product complexity.

The hardware engineers similarly had a considerable ability to reduce the complexity of their product; they achieved this ability not through assumptions, approximations, or reduced dimensions, as did the structural engineers, but through abstraction. Rather than conceptualize the chip in terms of the hundreds of thousands of logic gates that defined it at its most basic level, they conceptualized the chip in terms of information flow among components. Rather than talking in terms of transistors, voltages, and currents, they talked in terms of instructions and information that traveled through memories, multiplexers, and multi-gigabit transceivers. Their use of heightened abstraction reduced the number of components they had to worry about and enabled them to think holistically about their design.

The automotive engineers, unlike the structural and hardware engineers, had only minimal ability to reduce product complexity. Whereas beams and columns retain their shape along their entire length, automotive parts change shape along and across dimensions, at the bottom looking nothing like they do at the top, on one side looking nothing like they do on another. The asymmetry of automotive components and the vehicle as a whole made it difficult for automotive engineers to employ assumptions, approximations, or 2-D representations to reduce complexity. And whereas the engineers could group vehicle components into subsystems, the members of those subsystems were not identical in the way that logic gates are in microprocessors. For this reason, abstraction offered no solution either. Automotive engineers were uniquely forced to contend with the full complexity of their product.

Technology Factors. The final two occupational factors concern technology itself. *Technology cost* is simply the monetary cost of the technologies available for purchase by the engineers and their firms. In structural engineering, the cost of new technologies is rather low, with the most expensive software package at the time of our study weighing in at about $25,000 and many software programs acquired free or at nominal cost by engineers during the course of their university studies. In hardware engineering, by contrast, the cost of new technologies is high, with many applications costing upward of $1 million for a single license. In automotive engineering, the cost of the technology needed to maintain physical testing facilities—including racetracks, proving grounds, garages with experimental bays, wind tunnels, and the like—and of the associated personnel is indeed high. That high cost was exactly why the managers we observed hoped to replace physical tests with virtual tests. The cost of simulation technology, in comparison to the cost of physical testing facilities, is quite small. Maintaining both systems of testing thus incurs the full onslaught of technology costs.

The factor of *technology transparency* refers to the extent to which engineers can "peer into," or come to intimately understand, the inner workings of the technologies they employ. For the structural engineers we observed, the level of technology transparency was low. Although structural engineers could easily see the formulas embedded in Excel spreadsheet templates that their colleagues had created, the commercial software applications they employed were largely "black boxed." That is to say, structural engineers had no access to the code of commercial software applications, and the documentation that accompanied software typically failed to provide details of the algorithms, data structures, and decision rules that governed the technology. Structural engineers could supply data as input to these technologies and observe output, but they had limited visibility into how the technologies operated.

The opposite situation held true for the hardware engineers, whose technology transparency level was high. These engineers often coded their own scripts or borrowed the scripts of their colleagues; as a result, they had direct access to the inner workings of the software. In the case of commercial applications, they often worked hand in hand with vendors, in the process gaining high visibility into each new technology's operations. Even when they downloaded scripts from unknown persons on industry websites, they still had the ability to directly view—and alter—the code in those scripts. In short, hardware engineers rarely worked under conditions of not knowing what algorithms, data structures, and decision rules their technologies employed.

The automotive engineers' technology transparency was moderate on average, but in individual cases could be high or low. In an example of high transparency, although analysis engineers may have constructed input decks for their simulation models through the structured interface of a commercial application, when they wanted to modify that deck for a second, slightly different simulation, they simply opened the input deck directly in a text editor and altered the code they found there. Additionally, with an assortment of groups within the firm dedicated to technology development, the automotive engineers were never too distant from the creators of many of the firm's computer applications, which means that they may have gained access to or understanding of home-grown applications through these local creators. In the case of CrashLab, by contrast, which automatically generated input decks without allowing engineers to see what it had added to those decks, a new technology lowered transparency. Thus, this field gets a mixed, or moderate, rating for this factor.

Table 6.1, which summarizes how these occupational factors manifested across the three engineering occupations in our study, makes clear that, although the occupations were similar in type of work, type of technology, and product complexity, they otherwise differed in terms of how occupational factors manifested across them. With the exception of self-explanation of work artifacts, which was low in all three occupations we studied, each factor (reading along the rows of table 6.1) manifested differently across the three occupations. Additionally, no two occupations (reading down the columns of table 6.1) had a similar profile in terms of how occupational factors manifested in them. Given this variation across the three occupations, we next turn to and pull together our discussions in chapters 3, 4, and 5 to show clearly which occupational factors shaped technology choices by occupation and relationship.

Occupational Factors' Influence on Technology Choice by Occupation and Relationship

In chapters 3, 4, and 5, we described the technology choices that engineers in each occupation made. Our descriptions of engineers' choices included the motivation for their choices as well as the occupational factors that shaped motivations and subsequent choices. Tables 6.2 through 6.4 summarize the arguments from those three chapters for structural, hardware, and automotive engineering, respectively, pulling together our discussion across chapters (i.e., across relationships) so that we might form a holistic understanding of the factors at play in each occupation.

In each table, a check mark appears if a factor shaped a technology choice, as was true of *liability and government regulations* in structural engineering's choice to minimize the computer's role in the human-technology relationship, as shown in the corresponding cell in table 6.2. If no check mark appears in a cell, that factor played little or no role in shaping the corresponding choice, even though it may have facilitated that choice. Such was the case for *market* and *competition* in the human-technology choice of structural engineers: Custom products and bid-based competition certainly facilitated structural engineers' choice to minimize the computer's role (if only because they did not push for working faster, providing a good argument for computer use), but they did not cause the engineers to make this choice. Structural engineers could just as easily have worked on custom products with bid-based competition while maximizing the computer's role had they so chosen; other occupational factors, however, turned them against this choice. With this understanding of tables 6.2–6.4 in mind, we can consider each occupation in turn to review the factors that shaped its choices.

Structural Engineering. In structural engineering, a logic of understanding motivated the choice to minimize the computer's role (the first relationship column) and a desire to avoid errors motivated the choice to shun automated links (the second relationship column). Table 6.2 makes clear to us, when we compare these two columns, that the factors that influenced these two choices within structural engineering were nearly identical. Three factors—liability and government regulations, testing capability, and technology transparency—each played a role in these choices. The factor of liability and government regulations was high in structural engineering, as licensing examinations, professional stamps, and mandated peer review all attested. This factor strongly shaped which tasks structural engineers were willing to allocate to their technologies and how far they were willing to let their technologies operate together without their intervention. Structural engineering's low testing capability had a similar effect. Without an ability to test the soundness of their designs either physically or virtually, structural engineers were reluctant to allocate tasks to technologies or to permit a series of technologies to function without their intervention. The low transparency of technologies in this field compounded this effect by reducing structural engineers' ability to trust how technologies carried out work, prompting them to allocate fewer tasks to computers and to maintain a human interface between technologies.

The sole difference between the first two columns is the addition of "market" in the technology-technology relationship in the second column. Because structural engineers built unique products for individual clients, the paths that their solutions needed to take among their suite of technologies varied. Projects done in wood required a different path, for example, from that of projects done in steel, which in turn required a different path from that of projects done in concrete. Projects with mixed materials required yet another path. These differences rendered automated links between technologies difficult, thus suggesting market as a shaping influence in shunning these links.

The third relationship column, representing the choice about distant work, looks strikingly different from the first two columns; across the three columns, it features the largest number (five) of influencing factors: two

Table 6.2
Occupational Factors That Shaped Technology Choices in Structural Engineering

			Relationship		
			Human-Technology	Technology-Technology	Human-Human
Technology choice			Minimize computer's role	Shun automated links	Reject distant work
Motivation			Logic of understanding	Desire to avoid errors	Need to maintain base of domain knowledge
Occupational Factors	Environment	Market		✓	
		Competition			✓
		Liability and government regulations	✓	✓	✓
	Knowledge	Rate of knowledge change			✓
		Testing capability	✓	✓	✓
		Self-explanation of work artifacts			✓
	Work Org.	Division of labor			
		Task interdependence			
	Product	Product interdependence			
		Complexity reduction ability			
	Technology	Technology cost			
		Technology transparency	✓	✓	
Total			3	4	5

Note: A ✓ indicates that the factor strongly shaped the technology choice environment factors (competition plus liability and government regulations) and all three knowledge factors strongly shaped the engineers' choice to reject distant work. Competition influenced this choice because bid projects never included the luxury of overtime costs, thus providing an economic disincentive to allow engineers to work at home in the evenings or on weekends. Liability and government regulations caused senior engineers, whose professional stamps would mark the final solutions, to want their junior colleagues, whose calculations and models populated the stamped solutions, nearby so that they could easily monitor and advise all project work. The knowledge factors played a role because structural engineers could not fully test with any assurance the assumptions of their models, nor could they gain from examining past artifacts the design rationale that others took in similar projects. Rather, they needed to gain understanding from their seniors, whose vast stores of knowledge remained useful owing to the slow rate of knowledge change in this field. So great was this need to maintain the base of domain knowledge in the field that it provided the main motivation to reject distant work.

Hardware Engineering. Table 6.3 is remarkable at first glance in that it shows how different the three relationship columns for hardware engineering are as compared to the columns for structural engineering. In fact, the three factors that most strongly shaped technology choices in structural engineering—liability and government regulations, testing capability, and technology transparency—played no role at all in technology choices in hardware engineering. In hardware engineering, two factors, market and competition, and very nearly only these two factors, strongly shaped technology choices.

Getting to market with a mass product before other firms did was the key to success in this field, driving hardware engineers to work quickly. This logic of speed meant allocating as much work as possible to the computer, thereby maximizing its role. The desire to speed up work, combined with a desire to free engineers to work on coding tasks, prompted the embrace of automated links so that tasks ran even faster absent human intervention. Market and competition also prompted a need to coordinate, which motivated the engineers to collocate so that they could have their questions answered quickly and so that they could solicit detailed and involved collaboration at a moment's notice. The pressures that arose from market and competition in this field were largely unheard of in structural engineering (as the rows for market and competition indicate in table 6.2); in that field, firms bid on custom projects, operated according to an agreed-on schedule, and rarely allowed overtime.

Only two other factors, the self-explanation of work artifacts and product interdependence, shaped technology choices in hardware engineering, and each of them shaped only the choice in the human-human relationship. As in structural engineering, in hardware engineering the technology choice in the human-human relationship appeared to be the most tightly constrained; four factors shaped this choice, as compared to only two factors for the other two choices. The low self-explanation of artifacts meant that hardware engineers, like structural engineers, had difficulty gleaning design rationale from past design solutions; they needed the designers of those past solutions nearby so that they could ask them how and why the code worked as it did. In the case of hardware engineering, high product interdependence exacerbated the problem because new code relied heavily on old code, which meant designers had no option but to figure out how old code worked when writing new code. Thus, these two factors helped shape the choice to reject distant work.

Automotive Engineering. Unlike the tables for structural and hardware engineering, in which all factors with a check mark worked in the same direction to shape a technology choice, table 6.4 reflects in its first two columns sets of factors not all of which pulled in the same direction. We use capital letters—C, P, E, and S, described below—to help us distinguish in the table the direction in which each factor pulled.

In the first column, which reflects the human-technology relationship and the choice regarding the computer's role, we are reminded that a logic of belief motivated automotive engineers to achieve a balance between the computer's role and the role of physical parts and physical tests. The choice to balance the two roles meant that some occupational factors pulled in the direction of the computer (C), while others pulled in the direction of physical parts and physical tests (P). Specifically, market and competition both argued for speed (similar to the case in hardware engineering) and a reduction in costs, which favored the computer and pushed the choice in that direction. Technology cost, in particular the high cost of maintaining the proving grounds and other physical testing facilities, operated similarly. Pulling in the opposite direction, in favor of physical parts and physical tests, were liability and government regulations, as well as testing capability. Because government agencies carried out their own physical, not virtual, tests to ensure product compliance, automotive engineers maintained a retinue of physical tests. The fact that computer simulations so often failed on first try to match the results of these physical tests argued strongly in favor of maintaining physical testing facilities.

Table 6.3
Occupational Factors That Shaped Technology Choices in Hardware Engineering

			Relationship		
			Human-Technology	Technology-Technology	Human-Human
Technology choice			Maximize computer's role	Embrace automated links	Reject distant work
Motivation			Logic of speed	Desire to free engineers for coding tasks and to speed up work	Need to coordinate
Occupational Factors	Environment	Market	✓	✓	✓
		Competition	✓	✓	✓
		Liability and government regulations			
	Knowledge	Rate of knowledge change			
		Testing capability			
		Self-explanation of work artifacts			✓
	Work Org.	Division of labor			
		Task interdependence			
	Product	Product interdependence			✓
		Complexity reduction ability			
	Technology	Technology cost			
		Technology transparency			
Total			2	2	4

Note: A ✓ indicates that the factor strongly shaped the technology choice.

Table 6.4

Occupational Factors That Shaped Technology Choices in Automotive Engineering

			Relationship		
			Human-Technology	Technology-Technology	Human-Human
Technology choice			Balance computer's role with role of physical parts and physical tests	Embrace automated links (managers and development groups) Shun automated links (engineers)	Employ distant work
Motivation			Logic of belief	Desires for quality, speed, and standardization of practice	Need to reduce cost and time
Occupational Factors	Environment	Market	✓(C)	✓(E, S)	✓
		Competition	✓(C)	✓(E, S)	✓
		Liability and government regulations	✓(P)	✓(E)	
	Knowledge	Rate of knowledge change			
		Testing capability	✓(P)		
		Self-explanation of work artifacts			
	Work Org.	Division of labor		✓(S)	
		Task interdependence		✓(S)	
	Product	Product interdependence			
		Complexity reduction ability		✓(S)	
	Technology	Technology cost	✓(C)	✓(E)	
		Technology transparency		✓(S)	
Total			5	8	2

Notes: A ✓ indicates that the factor strongly shaped the technology choice. C and P refer to whether the factor argued for the computer or physical parts/physical tests, respectively. E and S refer to whether the factor argued for embracing or shunning automated links, respectively.

Similarly, in the second relationship column, we see the factors that shaped managers' choice to embrace automated links (E), to satisfy desires for quality, speed, and standardization of practice; we see as well the factors that shaped engineers' choice to shun automated links (S), based on their shared desire for speed but their differing perception of how to achieve it.

Occupational factors that argued in favor of embracing automated links included the three environment factors (market, competition, and liability and government regulations). The market and competition factors argued for speed, which automated links presumably offered. The liability and government regulations factor posed an argument for the forced standardization of practice, which automated links largely guaranteed. Technology cost, here in the form of the cost to maintain multiple licenses for a variety of different technologies, all of which performed the same test, argued for automated links to limit the number of technologies involved and thus reduce costs.

Factors that argued in favor of shunning automated links were the division of labor, task interdependence, and complexity reduction ability, all of which worked in concert to create a considerable number of engineering groups isomorphic to the product. Clear divides of task and technology interdependence separated the groups; these divides resisted automated links because a joining together of technologies would require a similar joining together of work roles or groups, a change that would considerably unhinge the division of labor at IAC. These factors thus served to limit managers' application of automated links to technologies within engineering groups, not across them.

One factor that shaped S&C engineers' shunning of automated links was technology transparency. The automated links in CrashLab made it impossible for engineers to "see" into input decks and forced engineers to return to the application's interface every time they wished to alter the input to construct a new variation of a test. The reason why market and competition spoke in favor of engineers' choice to shun automated links while also speaking in favor of managers' choice to embrace them is that these factors argued for speed. Managers thought automated links would yield speed; engineers thought the links hampered speed. Because speed was a commonly desired outcome, these differences in perception meant that the factor worked both ways.

In the third column of table 6.4, we see that market and competition were the only factors that drove automotive engineering to employ distant work. Again, these factors argued for reduced costs and time, which managers thought they would achieve via "twenty-four-hour engineering" across global centers to the India center.

Overall, a striking difference between the table for automotive engineering and the tables for structural and hardware engineering is the range of factors and categories that it covers. Whereas the factors that influenced technology choices in structural engineering lay almost completely in the categories of environment and knowledge and the factors for hardware engineering were centered in the category of environment, the factors for automotive engineering spanned all five categories, with the biggest groupings in environment, work organization, and technology.

Summary. If we compare tables 6.2–6.4, we see that each occupation had a distinct profile of factors that shaped its technology choices: the patterns of check marks in the columns in table 6.2, for example, are not replicated in tables 6.3 or 6.4. Thus, not only do different sets of factors shape technology choices for different relationships within a single occupation, different sets of factors further shape technology choices within relationships across occupations. The same set of factors did not shape, for example, the choice in the technology-technology relationship in each occupation. This finding reinforces our claim that occupational factors shape technology choices, and that those factors and their influence vary by occupation. Tables 6.2–6.4 are useful for summarizing the factors at play within each occupation and for each relationship. However, these tables do not easily afford an investigation of how influential each occupational factor was overall. For that we require a different mechanism.

The Extent of Influence of Each Occupational Factor

Table 6.5 enables the investigation of each factor's influence by merging information from tables 6.2–6.4 with information from table 6.1. We placed a check mark in a cell in table 6.5 if the corresponding instantiation of a factor (e.g., minimal testing capability) shaped the corresponding technology choice (e.g., minimize computer's role). Two check marks indicate that the instantiation of the factor shaped the choice in two occupations. In no case did an instantiation of a factor shape a choice in all three occupations.

Although our purpose in this book is primarily to establish that occupational factors shape technology choice, the arrangement of information in table 6.5 helps us push a bit further by tentatively exploring what our data might suggest for the universal effect, if any, of each factor. In particular, we are interested in several possible combinations of factors and choices. We want to know whether the influence of some factor seemed to

Table 6.5

Influence of Occupational Factors by Relationship and Technology Choice

Relationship								
Occupational Factors		Human-Technology		Technology-Technology		Human-Human		Total
	Technology choice:	Minimize computer's role	Maximize computer's role	Shun automated links	Embrace automated links	Reject distant work	Employ distant Work	
Market	Custom			✓				8
	Mass		✓✓	✓	✓✓	✓	✓	
Competition	Bid-based					✓		8
	Time-based		✓✓	✓	✓✓	✓	✓	
Liability and government regulations	Low							5
	High	✓✓		✓	✓	✓		
Rate of knowledge change	Slow					✓		1
	Moderate							
	Fast							
Testing capability	Minimal	✓		✓		✓		4
	Considerable	✓						
Self-explanation of work artifacts	Low					✓✓		2
	High							

Table 6.5
(continued)

Relationship							
Division of labor	Simple						1
	Complex			✓			
Task interdependence	Low						1
	High			✓			
Product interdependence	Low						1
	High					✓	
Complexity reduction ability	Minimal			✓			1
	Considerable						
Technology cost	Low						2
	High		✓		✓		
Technology transparency	Low	✓		✓			3
	Moderate			✓			
	High						

Notes: Each ✓ indicates that the factor strongly shaped the technology choice in an occupation; two such marks in a single cell indicate that it did so for two occupations. There are 37 check marks here but only 35 check marks tallied across tables 6.2 through 6.4 because we plotted market and competition for automotive engineers twice to match their dual influence.

follow a pattern, in that a low instantiation, say, shaped one choice (e.g., to embrace automated links) and a high one shaped its alternative (e.g., to shun automated links). That would speak to the potential predictive power of that factor, such that if we knew the instantiation of the factor for a given occupation not in our study, we might hazard an educated guess as to that occupation's technology choice. Likewise, we are curious to know whether the same instantiation of a factor shaped different choices across the occupations we studied or, more weakly, shaped a choice in one occupation but not another. Such a result would suggest that the factor has limited predictive (and hence deterministic) potential, with an impact that is dependent on its context. We are similarly interested to learn whether or not different instantiations of a factor shaped the same choice, again hinting at the limited predictive power of that factor. Finally, we contemplate why some factors had limited influence on technology choices across occupations. With these possibilities in mind, we examine the factors in turn.

Market and Competition. We discuss these two factors together because they seem to operate in a coordinated manner. We can say little about the influence of a custom market or bid-based competition; we had only the example of structural engineering in which they manifested, and the impact there was small. A mass market and time-based competition, however, both of which appear often in table 6.5, may bode for maximizing the computer's role and possibly embracing automated links between technologies, as attested by the double check marks in these cells. That is because a mass market and time-based competition argue for reduced time and costs, which, in the context of hardware and automotive engineering, the computer and automation offer in comparison to the human. In other words, if time and cost are paramount concerns, an occupation may be more likely to make technology choices that favor maximizing the computer's role and automating links across technologies. Across our three occupations and twelve factors, market and competition were by far the most influential factors, but it bears repeating that their influence occurred primarily in hardware and automotive engineering, not in structural engineering, where the market was custom and competition bid-based. In short, these factors strongly shaped choices in two of our occupations but not in the third.

Liability and Government Regulations. Low levels of liability and government regulations had little influence on the technology choices that we examined, other than that they (arguably) permitted the maximization of the computer's role, though they did not cause that choice. High levels of liability and government regulations, by contrast, argued strongly for minimizing the computer's role. In the case of structural engineering, the alternative to the computer was a human thinking through the model and its assumptions; in automotive engineering, the alternative was the physical test. In both cases the alternative emphasized a process that was more immediate to the human than the computer would have been. Beyond the computer's role, the effect of high levels of liability and government regulation is unclear. What our data do suggest is that, just as market and competition seemed to be influential only if they brought time and cost concerns into sharp relief, liability and government regulations appeared to be influential only when they were at high levels. In fact, we had the sense that, were the federal government ever to discontinue physical testing in favor of virtual testing, automotive managers, keen to cut costs and time, would be quick to follow suit. In other words, the absence of liability concerns and government regulations appears likely to leave technology choices wide open. Overall, because the factor of liability and government regulations was high in two of our three occupations, it proved to be one of the more influential factors in our study.

Rate of Knowledge Change. The rate of knowledge change had only one shaping effect on a technology choice: when the rate of knowledge change was low, this factor favored the rejection of distant work, so that junior engineers might be close to senior engineers from whom they might learn. Beyond that, this factor had no strong influence on technology choices across the three occupations we studied.

Testing Capability. Our findings suggest that minimal testing capability may be a force toward rejecting the computer on all the fronts we studied: minimizing its role, shunning automated links, and rejecting distant work. But having considerable testing capability did not guarantee a free path to those individuals who preferred the computer. In the case of automotive engineering, where considerable testing capability was reflected in the ability to validate simulation tests against physical ones, the engineers chose to limit the computer's role because too often the computer test was found to be wrong in its predictions. Overall, testing capability strongly shaped choices primarily in structural engineering, where it was low; this occupation accounted for the bulk of this factor's influence.

Self-explanation of Work Artifacts. The low self-explanation of artifacts spoke strongly for the rejection of distant work because individuals needed to be close to their colleagues to gain explanations of knowledge left implicit in the artifacts. This situation was characteristic of structural and hardware engineering, both of which occupations rejected distant work. The low self-explanation of artifacts argued for that same rejection of distant work and for the same reasons in automotive engineering (we recall here the difficulties the India center engineers had reading reference models), but managers overlooked that fact when making the choice to offshore.

Division of Labor, Task Interdependence, Product Interdependence, and Complexity Reduction Ability. We can say little about the potential predictive ability of any of these four factors because each one was influential in a technology choice only once. We do note that, in each case, the factor was influential in shaping a choice when it was at its "bad" extreme. That is to say, when the division of labor was complex (not simple), when task and product interdependence were high (not low), and when the ability to reduce complexity was minimal (not considerable), then these factors shaped a choice. When these factors were at their "good" extreme, all choices appeared to be open; in other words, these factors had no influence in shaping choices in those situations.

Technology Cost. Similarly, when technology costs were high in our study, they had more influence than when they were low. However, the overall influence was not that significant, with only two choices affected by technology cost. In both cases the choices were in automotive engineering. The high cost of physical testing facilities spoke to maximizing the computer's role; the high cost of maintaining multiple computer licenses spoke to automating links across technologies to enforce single options (and releasing other licenses). Interestingly, the cost of technologies proved not particularly influential in shaping engineers' choices about technology in our study. In comparison to market and competition factors, the low influence of technology cost suggests that, when it comes to economics, the key is not in the technology itself but in what larger choices the technology affords.

Technology Transparency. Finally, when technology transparency was at its "bad" value of low transparency, it shaped more choices than when it was either moderate or high. When the ability to see into computer technologies was low, the motivation to use them or to link them was low. Although technology transparency appears to have had a limited influence on technology choices, we found it an intriguing factor because it speaks to an individual's ability to intimately understand the technologies that aid his or her work, and we are curious what its broader potential may be across a wider range of occupations.

What We Learned about the Factors That Shaped Technology Choice

Our study of technology choices in three occupations makes clear that occupational factors shaped these choices. Our study also makes clear that different sets of factors were influential across occupations. The range of factors was broad, covering five categories: environment, knowledge, work organization, product, and technology. Across these five categories, a total of twelve factors were involved in shaping technology choices. These results have enabled us to establish "proof of concept," as it were, for our thesis that occupation matters when it comes to technology choices. With only three occupations and three technology choices, however, we could only tentatively explore the larger predictive role that any individual factor may have across occupations and choices. Based on that tentative exploration, four factors among the twelve that emerged from our data appear to have the largest potential. These four factors are market, competition, liability and government regulations, and testing capability. In addition, the self-explanation of work artifacts, when low, seems to have a highly consistent effect on the choice to reject distant work.

7 An Occupational Perspective

An occupational perspective on technology choices like the one we develop in this book has substantial benefits and implications for scholars, managers, practitioners, technology designers, and policy makers. Like all perspectives, an occupational perspective also has its share of disadvantages and raises some concerns, which we address in this chapter in terms of assumptions and limitations.

Benefits and Implications of an Occupational Perspective

Scholars. In chapter 1, we followed the work of the organizations scholars Paul Leonardi and Stephen Barley to untangle distinctions between determinism and voluntarism, on the one hand, and materialism and idealism on the other.[1] Scholars who explore technology choices too often ignore these distinctions, positing the two dominant perspectives on technology choice—technology determinism and social constructivism—as direct opposites. Our purpose in untangling these distinctions was to clearly situate alternative perspectives in relation to the two dominant ones. We reproduce in table 7.1 the table from chapter 1, with the inclusion now of our occupational perspective as an idealistically deterministic alternative (in the lower left quadrant of the table). Alternatives to the dominant perspectives are in the shaded cells of table 7.1.

As we discussed earlier, alternative perspectives such as sociomateriality and critical realism combine materialism with voluntarism in an attempt to explain technology choices and their effects by showing how the line between what is social and what is material blurs when one looks closely at it. Through philosophical discussions about the relationship between technology and organization, scholars show how these perspectives in materialistic voluntarism explain technology choices as the result of strategic actors negotiating or enacting the boundaries between the social and the

Table 7.1
Mapping Our Occupational Perspective on Technology Choices

	Determinism *Outcomes are inevitable; external forces are the agents of change*	Voluntarism *Outcomes are not inevitable; humans are the agents of change*
Materialism *Physical causes drive human action*	Technological determinism • Contingency theory • Fitts lists	Sociomateriality Critical realism
Idealism *Ideas and beliefs drive human action*	De/Upskilling theories Occupational perspective	Social constructivism • Structuration theory • Practice theory

material. Although these alternative perspectives are intriguing and have inspired considerable research about technology choices, they are limited in their ability to explain consistency in technology choices across contexts. Moreover, likely as a result of their strong grounding in the epistemology of voluntarism, sociomateriality and critical realism appear almost ideologically resistant to making predictions about how technology choices might unfold.

For this reason, we advocated further scrutiny of the undertheorized stance of idealistic determinism, and it was here that we built our occupational perspective on technology choice. Chapter 6 laid out in detail the findings of our exploration of the technology choices that structural, hardware, and automotive engineers made when they considered incorporating new technology. These choices, we found, concerned the relationships between humans and the new technology, between the new technology and existing technologies, and ultimately between people. We identified twelve occupational factors related to environment, knowledge, work organization, product, and technology that strongly shaped technology choice within occupations. Specifically, we found that market, competition, liability and government regulations, and testing capability were among the most important factors driving technology choice across the three occupations that we studied.

From a theoretical standpoint, what is perhaps most interesting about these findings is that these four factors produced remarkable homogeneity in technology choice across organizations by occupation. In other words, different organizations with distinct founding histories, clientele, and members made identical technology choices within the same occupation.

Such similarity is a sign that determinism may be at play. But this determinism is not a determinism of a materialist type because the three occupations that we studied all employed some form of sophisticated computational, graphical, and logical computer applications. The engineers in each firm had individual desktop computers, servers, and Internet connections. In short, they all had advanced information and communication technologies at their disposal, but this similarity in technology did not drive similar choices across occupations. Hence, materialistic determinism cannot explain our findings.

Rather, the determinism we detected in our study is of an idealist type. Idealists propose that shared ideas, beliefs, norms, and values drive human action. And, as many scholars have argued, there are few social enterprises that more carefully manage, regulate, and control ideas, beliefs, norms, and values than do occupations.[2] What our analysis has shown is that these occupational factors strongly shape technology choices within an occupation. Thus, scholars and others who wish to predict technology choices might do well to begin with occupations by mapping occupations' ideas, beliefs, norms, and values to technology choices, and in the process identify the factors with the most predictive potential.

Our occupational perspective transcends the debates between technological determinism and social constructivism by incorporating ideas from both perspectives. Like technological determinists, we believe that external forces shape outcomes. Like social constructivists, we believe that ideas and beliefs shape human action. We contend, in short, that individuals' ideas and beliefs at work with respect to the technology choices that face them derive from occupational factors. On the basis of these occupationally inspired ideas and beliefs, individuals make choices about technology. Therefore, we see similarity in technology choices—in the role that a technology plays, in its patterns of use, and in its consequences—within occupations, and differences in choices between occupations.

As a result, an occupational perspective can account for why technology choices are similar across organizations without claiming that they have to be similar across all occupations. Similarly, it can account for why technology choices sometimes differ. In short, an occupational perspective frees scholars from materialistic determinists' claim that technology choices are universal, and from the tendency of social constructivists to suggest that technology choices are specific to the local context in which they arise. An occupational perspective argues that technology choices are likely to be far-reaching and generalizable, but only to the boundaries of a particular occupation. Rather than claiming that every organization will make similar technology choices or that every organization's choices will be different, an occupational perspective suggests that many choices will look the same, and it explains why and under what conditions such similarity is likely to occur.

Finally, our work makes a particular contribution outside the boundaries of research on technology choices. That contribution is to studies of institutionalism. Institutionalists have long held that occupations aid in the diffusion of practices across organizations and communities because occupations represent strong social collectives whose members travel across many different workplace organizations. Because members within the same occupation share similar ideas, beliefs, norms, and values, the migration of an occupation's members across organizations tends to enhance homogeneity of practice.[3] Although institutionalists revere and hold as true this idealist stance concerning occupation-led institutionalization, they have scant empirical data to explain which occupational factors shape the spread of practices and why these factors shape practices as they do. Our data and the analyses in this book help fill this conceptual gap. We identified which occupational factors shape the choices that organizations make about how to implement and use new technologies, and we showed how these factors do so commonly within occupations. By choosing three occupations within the broader domain of engineering, we were also able to demonstrate that the boundaries around occupations are strong. That is to say, although structural, hardware, and automotive engineers perform similar types of design and analysis work, employ similar technologies, and work on similarly complex products, they make dramatically different technology choices.

Managers and Practitioners. For managers and practitioners, an occupational perspective can aid in thinking through technology choices, past, present, and future. To begin, an occupational perspective reveals to managers and practitioners where relevant comparisons lie. An occupational perspective tells managers and practitioners that, when trying to ascertain the potential impact of automation, say, or new visualization technologies or new communication technologies prior to purchase, they should consider cases from their own occupation, as opposed to examples from other occupations, as most appropriate for assessing their own situation. These cases would provide the most relevant comparative information because an occupational perspective suggests that the role a technology will come to play, the patterns of use it will prompt, and the consequences of that use are likely to be more similar within an occupation than across occupations.

More broadly, an occupational perspective can guide managers and practitioners in sorting through the historical, social, economic, and political factors that are likely to shape their technology choices. The structural engineers we observed, at least the senior ones, appeared to understand very well the factors that shaped their choices. They resisted the lure of new technology—certainly a difficult endeavor when working in a technical occupation—because they understood and accepted the factors that shaped their choice to minimize the computer's role. Junior engineers, fresh out of university programs in which computer applications reigned supreme, challenged, though weakly, the senior engineers' technology choices. Some junior engineers believed their senior colleagues were simply old-fashioned, afraid of change, or inept with new technologies. It was a matter of some curiosity and speculation on our research team, on first observing this belief among some junior engineers, what we might find in the field twenty years hence when the current generation of senior engineers will have retired and the current junior engineers will be running the firms. The attitude of mid-level engineers, however, suggested to us that, over time, fresh engineers come to understand the worth of approximate analytical methods over flashy computer applications. Should that pattern persist, then computers may be no more prevalent in this field in twenty years than they are today. We return to the issue of technological advance in the limitations section of this chapter; the point here is that an occupational perspective can bring home to managers and practitioners the realization that occupational factors—and not the continuous fare of new technology options that vendors bring forth—are primarily what shape, and what ought to shape, technology choices.

Although the structural engineers showed an awareness of the factors that shaped their technology choices, the automotive engineers did not. Many managers and practitioners in this field seemed prone to being swayed by technological advances and the potential gains those advances held. For example, on viewing the iconic graphical representations of sophisticated simulations of vehicle performance, managers came rather quickly to the conclusion that physical testing facilities were no longer required. Based on this erroneous conclusion, they set up the digital engineering services center in Bangalore without an accompanying proving grounds, a testing garage with experimental bays, or tear-down rooms, and they populated the center with engineers who had little personal familiarity with vehicles. An occupational perspective might help managers and practitioners resist some factors that work to shape their technology choices (here, market pressures that argue for substituting expensive engineering labor with its less expensive counterpart) by paying attention to other factors that equally affect their choices (here, the low self-explanation of work artifacts that made working with their distant creators difficult, among other factors).

An occupational perspective should further help managers and practitioners consider what the consequences of choices may be and how existing workplace phenomena may be the result of past technology choices. We noted earlier that the hardware engineers we interviewed had a freewheeling workplace culture in which strolling in to the office at 10 a.m. and playing foosball in the break room in the middle of the afternoon were perfectly acceptable acts not because the hardware engineers were representative of the hip, "new economy" mindset of Silicon Valley in the early 2000s but because their choice to maximize the computer's role in their work enabled them always to have a computer working for them even when they were not working on a computer. They were not goofing off, they were multitasking. Moreover, many of them were up in the middle of the night, working from home computers. In that same period of time in Silicon Valley the structural engineers employed in workplaces where foosball was unheard of and standard office hours were the norm were not simply indicative of a staid, "old economy" engineering that had yet to evolve. Rather, the structural engineers adopted those work habits because the engineers had minimized the role of the computer in their work. If the structural engineers were not sitting at their desk working, work was not getting done. In short, an occupational perspective cautions managers and practitioners who may be thinking about making changes to their workplace culture or lobbying for changes in work policies to first consider to what extent prior technology choices have led to current workplace culture and habits.

Technology Designers. For technology designers and funders of technological advance, an occupational perspective on technology choices can suggest occupations toward which designers and funders might profitably target new technologies, as well as what features those technologies should have. The designers of technologies in hardware engineering knew this fact well. In their domain, the engineers' near obsessive attention to technology meant that these designers were flooded with suggestions and requests for new features, as well as with advice on how to structure input and output interfaces. Technology designers in this field did not need to watch hardware engineers at work to understand technologies' role in the design process because hardware engineers were vocal in telling designers how technologies currently did, and optimally should, fit their design and analysis tasks. The hardware engineers conveyed this information to the technology designers at design automation conferences, at in-company technology marketing presentations, in the course of discussions held as computer chip firms served as test sites for beta technologies, in the extensive prepurchase investigation that hardware engineers conducted, and through the online conversations that hardware engineers held with each other across firms in technology user support groups and other forums. Because the hardware engineers chose to maximize the computer's role in their work, they were proactive in working with the technology designers to get the technologies they wanted and in the form they wanted them.

The designers of technologies in structural engineering did not enjoy the luxury of detailed information that their brethren in hardware engineering technology design did. And whereas that lack of information may have disturbed designers who wished to develop the best applications possible and frustrated users who had to contend with systems that did not pair well with their work practices, the fact that structural engineers minimized the role of the computer in their work meant there was little incentive to remedy the problem by developing a tighter relationship between users and designers in this field. Structural engineers, unlike hardware engineers, could afford to just get by with technology that was not perfectly attuned to their design and analysis tasks.

As a result, designers anxious to incorporate new technology advances into their structural engineering applications may find they need to pursue a different market. For example, we noted earlier that structural engineers had little need for 3-D visualization capabilities in their work and that, during our study, academic researchers in civil engineering were developing 4-D technologies that added the dimension of time to computer simulations. Although the designers intended the 4-D simulations to aid all participants in the architecture-engineering-construction triad, these technologies found a home not among structural engineers but among contractors and construction firms, which used the simulations to help them manage the activities of countless vendors and firms in the course of large construction projects.[4] Structural engineers, who saw little need for 3-D, saw even less need for 4-D.

Similarly, during the period of our study one senior structural engineer was enthusiastic about a knowledge management application that an academic researcher he knew had developed and that he was testing for her in his firm. He showed us how senior engineers could populate the application with explanations for various standard design solutions, for which junior engineers could then search when they had questions. The application failed to take off in his firm, in part for all the normal reasons that knowledge management attempts typically fail, including the difficulty of capturing all the knowledge that some individuals had and others needed. In addition, in the structural engineering world, in which all hours are billable, few senior engineers had the time to enter design solutions and rationales. Moreover, the junior engineers whom we queried about the knowledge management system reported that they did not use it because the system would have made public within the firm their query, and they did not want their ignorance revealed to everyone.

Perhaps more important, as our observations documented, only half the learning opportunities in structural engineering came about as a result of someone seeking knowledge; the other half came about as a result of someone offering knowledge. Senior engineers embraced their role as master teachers, and the teaching they offered arose in the course of everyday actions as they reviewed with their junior colleagues design solutions and ideas. Few openings existed in this intimate, side-by-side interaction for digital technology to play a role beyond displaying an image of a model or a design, and the engineers almost always preferred to stand together over large paper drawings rather than squeeze in front of small screens that could hold at one time but a fraction of the image. In fact, a look at the literature on knowledge management systems in structural engineering reveals just a few cursory attempts and little continued pursuit of this agenda.[5]

These examples illustrate the potential for an occupational perspective not only to inform new technology design and direct developers away from weak markets but also to direct the funding streams of agencies that invest in technology development. There is little point in developing high-end technologies for occupations in which existing technology choices suggest that new technologies are unlikely to gain great acceptance. Rather, with the understanding gleaned from an occupational perspective, funding agencies might focus their resources on occupations that stand to gain substantially from technological advances.

Policy Makers. For policy makers, an occupational perspective stands to deepen insights and target areas for action. Most immediately, an occupational perspective in the context of our current study has implications for the policy goal of ensuring an adequate supply of U.S. engineers, a topic that has captured national attention in the past decade. The goal of ensuring sufficient U.S. engineers involves policies on engineering education and visas for foreign-born engineers; it also draws into its discussion the offshoring of technical work. In fact, the crux of this policy debate lies in the potential global redistribution of the engineering workforce made possible by the annual graduation of thousands of engineers in countries such as China and India, engineers who are part of what the State University of

New York's Levin Institute terms the "global talent pool."[6] Fears that the United States is not keeping pace with rising economic powerhouses such as China and India and that the United States will soon lose its technical edge to other countries as engineering jobs dry up in the United States and move elsewhere drive the desire to avoid a shortage of U.S. engineers.

For this reason, policy discussions on the future of engineering in the United States have focused to date on numbers of engineers. Vagaries in definitions of who counts as an engineer and discrepancies in methods for counting engineers have led pundits to debate how many engineers the United States produces, how many engineers other countries produce, how many engineers are required in the United States, and how many U.S. engineering jobs are being lost to overseas competitors.[7] For example, Paul Otellini, former president and CEO of Intel Corporation, put forward a numbers argument in a 2011 *Washington Post* editorial.[8] Otellini pointed out that whereas China and India graduate a combined million engineers a year, the United States graduates but 120,000. U.S. firms that cannot fill engineering positions, he argued, will be forced to look elsewhere; moreover, without a sufficiently large engineering workforce, the United States will stop spawning innovative companies like Intel, the kind that spurs economic growth and provides countless jobs of all types.

An occupational perspective like the one we present permits a push past the limits of a numbers-focused debate. It does so by pointing out which engineering occupations may be most susceptible to wholesale offshoring of operations; which may tolerate a mixed model, in which labor is split between domestic and foreign subgroups; and which may see no offshoring whatsoever. For example, if structural engineering firms were unable to meet their labor needs with U.S. engineers, they might decide to move their entire operations overseas. They would have to make that choice because, faced with liability concerns and government safety regulations, a limited ability to test assumptions, and low transparency of technologies, senior structural engineers need to work side by side with junior engineers to make sure the domain knowledge, expertise, and intuition that the work demands are passed from one generation of engineers to the next. Hardware engineers, faced with a need to collaborate face-to-face when writing and debugging component code, might make a similar choice should they find insufficient U.S. engineers to hire. Other occupations, such as automotive engineering, might reckon that the combination of rationalized work and digitized work artifacts permits them to maintain disparate groups of engineers around the globe. In short, an occupational perspective helps us see that how offshoring will play out is likely to vary by occupation.

Moreover, an occupational perspective makes clear that shortages in engineering labor in the United States affect not only jobs for new or future graduates but also, potentially, current jobs. In some occupations, if not all engineering positions can be filled in the United States, it is likely that none of them will be. Thus, an occupational perspective also makes clear the courses of action for policy makers: namely, tailor government efforts—such as increased funding for research and grants for education—based on assessments of which occupations are facing shortages that threaten future or current American jobs. Making such decisions without an informed understanding of the impact of shortages invites trouble, as when, in the past, federal budget enhancements that increased PhD production in certain fields without a corresponding increase in PhD jobs in those fields contributed to a mismatch between supply and demand.[9]

We are not the first to suggest that numbers are insufficient for understanding and predicting the future of American engineering jobs. Scholars at Duke University, for example, have argued that the debate should also take into account the quality of engineering graduates.[10] When one removes from consideration Indian and Chinese engineering graduates, whose preparation in second-tier universities and colleges is substandard, then the situation looks less dire: American engineers are not competing with all of China and India's graduates, just the thin layer at the top. These scholars point to a 2005 McKinsey Global Institute survey of human resources professionals from eighty-three companies with worldwide operations whose results indicated that 80 percent of U.S. engineers were globally employable, whereas only 10 percent of Chinese and 25 percent of Indian engineers were.[11] Similarly, the international business scholars Stephan Manning, Silvia Massini, and Arie Lewin have argued that a numbers-focused debate ignores macroeconomic, policy, industry-level, and firm-level trends that shape offshoring decisions.[12] These trends include the availability of advanced information and communication technologies that make it possible to share work artifacts; the standardization and modularization of business operations, which make it possible to fine-tune the division of labor and facilitate task handoffs; cost pressures, which drive companies to lower labor costs; enhanced organizational capabilities in managing complex sourcing, which allow firms to distribute work; the increased demand for talent that growing economies spawn; and the rise of geographic clusters of skilled workers available for hire in emerging economies.

Our approach differs from approaches that focus on quality or large-scale trends in that those approaches have thus far largely treated all engineers alike. The definition of what constitutes a globally employable engineer, for example, does not vary by engineering occupation but instead refers to general attributes, such as the ability to speak English and to work on a team. Similarly, arguments about modularization presume work in all occupations is equally susceptible to this kind of division of labor; arguments about cost pressures assume all engineering occupations experience the same market-driven demands.

By contrast, an occupational perspective would suggest that the traits that make for an employable engineer vary by field. For example, a talented structural engineer is one who, when junior, is open to learning from seniors and, when senior, is capable of teaching juniors. These skills are not needed, and arguably frowned upon, in other engineering occupations such as hardware engineering. Similarly, when considering large-scale trends, scholars who discuss diffusing knowledge across far-flung engineering groups infrequently delve into the details of everyday work practices as we did. Instead, they restrict their concerns to knowledge-management systems and incentive structures for sharing information, as if all engineers worked in the same manner, under the same conditions, and subject to the same constraints. These discussions fail to appreciate, for example, that although hardware engineers do seek knowledge from their colleagues, they need to acquire it face-to-face, side by side, and in the moment, staring with a colleague at a computer screen while puzzling out what the code is doing. Discussions of engineering shortages that fail to distinguish among the work practices in occupations can yield only general advice about offshoring that is unlikely to speak well to all manner of engineers.

We should also offer the reminder that when we talk about an occupational perspective, we do not construe occupations as broad categories, such as doctoring, lawyering, or engineering, but as particular disciplines within those broad categories. Thus, in this book we have investigated not the occupation of engineer writ large but the particular occupations of structural engineer, hardware engineer, and automotive engineer. In the offshoring literature, adopting an occupational approach has typically meant simply making predictions about broad categories of workers. The regional economists Ashok Deo Bardhan and Cynthia Kroll, for example, took an occupational approach when they listed a set of attributes that they claimed would characterize an occupation at risk of offshoring (e.g., no face-to-face customer servicing requirement, high information content, a work process that is "telecommutable" and Internet-enabled).[13] Arguing that the wage differential between foreign and domestic workers was a pivotal attribute among this set, they claimed that occupations such as "computer and math professionals," "business and financial support services," and "diagnostic support services" were at high risk because foreign workers could do these jobs for substantially less pay than U.S. workers could. The occupational categories that these authors employed were obviously quite broad, which raises concerns about their accuracy.

Alan Blinder, a Princeton economist, came much closer to the level of detail that we favor when discretizing among occupations.[14] In a methodology that is as detailed in its investigation and analysis of compiled labor data as ours is with respect to empirical observational data, Blinder employed O*NET (the successor to the *Dictionary of Occupational Titles*) to make subjective assessments of the "offshorability" of various occupations. Blinder made his assessments based on his reading of O*NET descriptions of job titles, tasks, work activities, and the like for each of 817 occupations, then used this information to answer questions about the necessary proximity to U.S. work locations and units. The answers to these questions guided Blinder in placing each occupation in one of four offshorability categories that he devised: non-offshorable, hard to offshore, offshorable, and highly offshorable.

Occupations such as doctor, child-care worker, and amusement park attendant fell into Blinder's non-offshorable category because these individuals were tied to a specific work location (where the patient lived, where the child lived, where the park was). Occupations in which workers did not need to be close to others in their work unit fell into the highly offshorable category; here Blinder placed computer programmers, reservation agents, and mathematicians. If workers in an occupation needed to be close to one another and their entire unit needed to be in the United States (just not in a specific location), as would be the case for, say, radio and TV announcers, their occupation fell into the hard-to-offshore category. And if workers needed to be close to their work unit but the work unit could be abroad, as in the case of factory workers, the occupation fell into the offshorable category.

Blinder's method is painstaking because it requires human judgment of lengthy, detailed written attributes of each occupation; his application of this method to 817 occupations is singularly impressive. Although his approach is reminiscent of task-technology fit approaches in that it pays attention to attributes of the task in making a judgment about offshorability, it goes beyond, say, Levy and Murnane's fit approach because it does not ask, "Can this work be routinized such that a computer or a low-skilled foreign worker do it?" Instead, Blinder tries to take a broader view by asking questions about workers' proximity to locations and to others, and he does not fall into the trap of equating foreign workers with lesser-skilled workers.

Yet Blinder's approach is insufficient in part for reasons like those that troubled Levy and Murnane's approach, and Fitts lists before it. For Blinder's approach takes into account the host of historical, social, economic, and political factors that shape occupational choices only to the extent that the generators of O*NET's data did when writing their descriptions of workers' tasks, technology, knowledge, skills, and the like for each occupation. Occupation analysts provided O*NET's original data, which are updated annually through surveys of practitioners and occupation experts. What these informants respond to, however, is a standardized hierarchical model with 277 descriptors. Although 277 descriptors may sound like a lot of data, and they are, those data may not be the necessary data when considering offshoring choices.

As an example, structural engineers do not have their own O*NET category, instead falling under the broader category of civil engineers. Blinder placed civil engineers in his non-offshorable category on the ground that a person in that occupation needs to be physically close to a specific U.S. work location. Though Blinder did not spell out his decision process in the case of civil engineers, a reading of the O*NET data for this occupation suggests that he gave them this rating because civil engineers need to be close to the architects and clients with whom they work and the sites whose structures they are designing. While the former is more true than the latter (one can fly to the site on the rare times that doing so is needed, as our structural engineers did), that situation holds only for those senior engineers who meet with architects, clients, and contractors. The reason why the work of junior engineers is not offshored has nothing to do with proximity to architects, clients, or sites and everything to do with liability concerns and government safety regulations, the inability to fully test assumptions, and the lack of transparency in technologies. Only the first factor in this list, liability concerns and government safety regulations, does O*NET data consider, and not in the kind of depth that would make apparent why junior engineers need to be colocated with senior ones.

Nonetheless, Blinder's approach is intriguing, and it is the kind of approach that might be possible if the underlying data were complete. Currently, they are not, and that they ever might be is a matter of debate. To understand why, consider that in our study we have identified seven among twelve possible factors across three engineering occupations that shaped the choice of whether or not to offshore work. The O*NET data do not address these seven factors. Moreover, further studies of other occupations might yield yet more factors that shape this choice in other fields, requiring additional descriptors in the O*NET database. Consequently, an occupational perspective like the one we take in this book, with its attention to the details of everyday work practice as observed in the field and its rooting out of the factors that shape technology choices, provides policy makers with the best means for determining exactly which occupations are at risk of losing jobs to overseas workers.

Beyond issues of the engineering workforce, an occupational perspective has implications for other types of policy for a variety of occupations; we briefly note a few of them here. An occupational perspective stands to inform educational policy by highlighting how advances in technology may or may not change the knowledge, skills, and training that workers in a given occupation require. It could, therefore, guide funding and curriculum reform decisions in K–12 and post-secondary education. An occupational perspective would further help education policy makers think about which occupations—and hence which educational areas—are more likely to benefit from efforts to increase the supply of related talent and which are not. Although Erik Brynjolfsson and Andrew McAfee, scholars who study information technology and the economy, argue against implementing government policies that would restrict new technologies (even though these authors show convincingly how new technologies are dramatically reshaping the workforce and erasing opportunities for many workers), we think an occupational perspective could support the creation of "workforce impact reports" similar to environmental impact reports that would help policy makers in labor and technology. As the authors themselves note, "Technology creates possibilities and potential, but ultimately, the future we get will depend on the choices we make."[15] We agree that choices stand to be made, and would only add that occupational factors will strongly shape them. Thus, policy makers need to study those factors. Finally, as we have noted, an occupational perspective stands to inform national technology development policy by helping to identify occupations in which new technologies might significantly aid or transform work, and other occupations in which technological advances might be much less needed or desired. Such information should help decision makers who allocate budgets within funding agencies in research and development.

Assumptions and Limitations of an Occupational Perspective

Despite its benefits for scholars, managers, practitioners, technology designers, and policy makers, an occupational perspective comes with its own disadvantages and raises some concerns, which we address here in terms of assumptions and limitations. The first assumption we explore is that technology choices will not converge over time to some optimal, universal choice. The second assumption we address is our declaration of what counts as an occupational factor, as opposed to, say, an industry factor. We also discuss three limitations that, although particular to our study, are characteristic of the occupational perspective that we put forward: (1) our data were rooted in a particular period of time, and thus are prone to discounting by technological advances, (2) our data, because they were gleaned from a small sample, are constrained in their generalizability, and (3) our data required too many resources for their collection and analysis to make this approach realistic for future scholars.

Assumptions

Nonconvergence of Technology Choices over Time. An occupational perspective assumes that technology choices will not converge over time to a single "best" choice. Although our use of the term "choice" suggests a single event at a single point in time, is it not true that the role that technology plays in the workplace, its patterns of use, and its consequences evolve over time? How can we talk about a single, unchanging choice? In fact, how do scholars know that the social constructivists' argument, true perhaps in the short term, does not lose out to the technological determinists' argument in the long term?[16] That is to say, how do scholars know that, whereas the local context might shape initial use and transition to a new technology, eventually the affordances of the technology, which are universal, reveal themselves, causing all users to converge to a single pattern of use, leading to a single role for the technology and a single set of consequences that is invariant across sites?

Scholars have considered the role of time when looking at technology choices in the workplace, and their investigations reveal that patterns of use do develop over time. A good example of work in this realm is research by Marie-Claude Boudreau and Daniel Robey, who examined what happened in the wake of the introduction of an enterprise resource planning (ERP) system in a state government agency.[17] Boudreau and Robey found that at first, employees rejected full-scale use of the new system. For example, although part of the intention in adopting the system was to create a paperless office, employees resisted the move to digital-only record keeping by continuing to fill out paper forms and by printing forms they had completed on the ERP system, in effect creating a shadow system on paper that complemented the new digital system. Soon, however, employees came to realize that they would never reach the efficiency levels that managers expected of them if they continued in their paper-based ways. Slowly, they began to adjust to the new system. However, they found its interface and processes at times too restrictive. As a result, they developed work-arounds, invoking patterns of use unintended by the system's designers.

Organizational studies scholars Marcie Tyre and Wanda Orlikowski described the period right after a technology's implementation as a window of opportunity during which users can alter the technology, within the limits of its material flexibility and the limits of their context, to meet their needs, just as the state employees did in the case of the ERP system that Boudreau and Robey studied.[18] By the time that the window has passed, use has become routinized into certain patterns, which, shaped by local objectives, local situations, and local users, are likely to be unique to a given context. According to the social constructivists' argument, the local role that a technology plays, local patterns of use, and local consequences become set by the end of the window of opportunity. These outcomes do not converge to some optimum.

We agree that roles, patterns of use, and consequences of those uses are likely to be fluid directly after implementation of a technology as users come to know the technology and determine how it will fit into their environment, and to solidify after this window of opportunity for adjustment passes; the scholarly literature is fairly consistent on this point. In our study, however, we did not look at the implementation of a single technology. Rather, we looked at a suite of technologies and the choices that engineers made with respect to that suite. We did not arrive on the scene just before or just after engineers put their technology suite in place. Instead, we arrived on the scene to find well-established technology suites and their associated roles, patterns of use, and consequences. Certainly, the engineers did purchase new technologies during our time in the field. But those purchases were incremental additions to existing technology suites. The larger technology choices had already been made prior to our arrival, and the factors that shaped those choices also shaped new technology purchases. When we talk about choices, we are talking about the configurations that emerged after numerous windows of opportunity had come and gone; we are talking about larger patterns of behavior and belief.

Moreover, as we have demonstrated throughout this book, the strongest forces that shaped technology's role, patterns of use, and consequences in the three engineering occupations we studied were occupational ones, not local ones. Occupational forces shaped engineers' choice to adopt new technologies in the first place, they shaped how engineers appropriated those technologies to their design and analysis process, and they shaped the role that the technologies would take on, the patterns of use that arose, and the consequences of that use. Ultimately, technological determinism lost out in our study because roles, patterns of use, and consequences did not converge, as our data bear witness through the variation we saw across occupations. Likewise, the social constructivist argument for purely local outcomes also came under question in our study because of the similarity within occupations but across firms that we observed. Our study points to the potential of an occupational perspective that is unlikely to be challenged by convergence of use because that convergence was not evident in our study despite the presence of desktop computers and their many applications in engineering for more than two decades.

What Counts as An Occupational Factor. Some readers may be troubled by our inclusion of several factors deemed occupational that may lie beyond occupation. Are we incorrectly assuming that these factors belong in the occupational fold, and, as a result, are we masking influences that extend beyond occupation? Some factors, such as the rate of knowledge change, seem to lie comfortably within the confines of an occupational perspective. Most of us understand and accept that occupations rest on unique bodies of knowledge that speak to their expertise; we find it natural that knowledge and occupation coincide. But what about other factors? Let us consider, for example, environmental factors. Their very name suggests they may extend beyond the occupation. We do not deny that claim. For example, in the case of hardware engineering, the market was one of mass production not just for the hardware engineers in the firms but for the firms' software engineers, sales staff, human resources group, financial staff, and others. In that case, we do not mind that market was bigger than occupation; we care only that market was true for everyone in the occupation across the three firms that we studied.

We would be more concerned if market varied within an occupation. Are there hardware engineers, for example, who design custom one-off chips, chips for which only a single instantiation, not thousands or millions of instantiations, will be built? No doubt there are such hardware engineers out there working on highly specialized projects, but we suspect there are not many of them as compared to the mass of hardware engineers in the occupation. We allow, however, that there may be occupations in which market varies significantly among its practitioners. In those cases, one would need to fine-tune the market factor or find some replacement for it. Alternatively, one might need to draw finer lines to differentiate among occupations (a move that would preserve an occupational perspective) or possibly consider a different level of analysis (a move that would veer from an occupational perspective toward some other categorization). All we can say at this point is that the twelve factors that emerged from our data were consistent across the engineers and firms in our sample within each occupation. For us, these factors held as occupational factors, and that is why we deemed them such.

Limitations

Data Rooted in a Particular Time Period. Advances in technology have more than once put the lie to claims that scholars have made about technology. We studied engineers in a particular period in time, largely the first decade of this century, and the technologies at hand at this time. How do we know with any confidence that, in the years to come, advances in technology will not alter technology choices among engineers in the occupations we studied, and in so doing erase all support for our claims?

Let us assume, for the sake of argument, that advances in technology would not cause the engineers in our study to revert back to situations in which their use of technology becomes less than what it is now. That is to say, let us assume that technological advances would not prompt a move toward a smaller role for technology in an engineer's work, to a greater allocation to man over machine. If we accept this premise, then, by asking about the effect of technological advance, we are most concerned about those engineers who currently have, for what we have deemed purely occupational reasons, a somewhat limited or restricted embrace of technology. The question then becomes, Might that embrace not grow stronger with better, more capable, more advanced technologies?

The answer is a qualified yes, depending on exactly what features of the technology are advanced. Here we may consider structural engineering, where engineers minimized the role of the computer in their work because liability concerns and government safety regulations were paramount, because engineers had no good way to test whether their own modeling assumptions were true or not, and because the computer applications they had available to them were not transparent (that is to say, engineers had very little idea what algorithms the applications employed, what decision rules they followed, and the like). In this environment, if a new technology came along that provided better perspectives (say, 3-D or 4-D renderings), that plotted results in a more user-friendly format, or that performed advanced computations not included in the current technology options, that new technology would fare no better in the structural engineers' judgment than current options because the absence of those features is not what currently limits the engineers' technology choices.

If, on the other hand, a new technology came along that was fully transparent, whose documentation clearly explained its algorithms, processes, and calculations, whose code was written in a language approximating English so that engineers could follow along step by step, and whose interface allowed substitution and alteration at the engineer's discretion (e.g., skipping a computational step if so desired), then that technology might indeed alter the man-machine ratio in this occupation. Similarly, if a new technology could somehow guarantee full adherence to the state building code and freedom from liability claims, structural engineers might embrace it. In other words, not any advance has promise in this field, only advances that address the factors that constrain engineers' technology choices.

As evidence of this claim, we point to a study done by innovation scholars Mark Dodgson, David Gann, and Ammon Salter.[19] These scholars investigated the emergence of a new occupation, fire engineering, which arose in the wake of the events of September 11, 2001, and was made possible by new fire simulation technologies. These new technologies modeled the performance of structures on fire, the behavior of people evacuating buildings in emergencies, and the flow of smoke through burning structures. Employing advanced FEA and CFD algorithms as well as sophisticated simulations of crowd behavior, these technologies were state-of-the-art in the years directly after 9/11. The field of structural engineering largely spawned fire engineering, and these new technologies were coming onto the scene just as we had completed our fieldwork in structural engineering. Thus, fire engineering makes a good test case for the premise that technological advances will overturn our claims.

As Dodgson, Gann, and Salter reported, fire engineers, like their forebears in structural engineering, relied heavily on their experience and knowledge to guide their modeling efforts despite being armed with the latest in technological advances. They routinely returned to first principles, often working out designs via pencil drawings and hand calculations using standard engineering formulas. The task of modeling buildings on fire, as one engineer noted, did not entail "sticking things into models." Rather, because fire engineers had no good way to test their assumptions, their task involved iterating between real cases of fire and models of fire. In addition, fire engineers sought direct and immediate input from their colleagues in the office. Old-fashioned engineering knowledge, paper-based modeling, and face-to-face dialogue were essential to the proper use of the new fire simulation technologies, which engineers could not simply charge with the task of modeling buildings on fire. As one engineer in the study noted,

"There are dangers here, because some of the tools are readily available and downloadable from Websites. Anyone can use them. People can run their own fire simulations on whatever design they like. Sometimes they don't understand the assumptions in these models or the consequences of the results." In short, the case of new simulation technologies in fire engineering supports our argument that technological advances will not alter technology choices if they do not address the factors that constrain those choices.

Overall, we argue that advances in technology are unlikely to negate our argument. Rather, advances in technology will only change the outcomes that we observed if they directly pertain to and in some way alter the factors that we have discussed. By extension, we would also expect changes in the outcomes that we found in this study if the factors themselves changed. For example, if liability concerns and government safety regulations dwindle for whatever reason in structural engineering (perhaps because of legal changes or modifications of the building code), we might see a shift in how many tasks engineers in this field are willing to allocate to the computer. In all cases, what reigns over technology choices are occupational factors.

Data Gleaned from a Small Sample. The second limitation concerns the size of our sample. Although a sample that spans three occupations, seven firms, eight countries, and fourteen sites (see the appendix for full details) is immense by qualitative standards, it is small in comparison to the full population of occupations, firms, countries, and sites that exist. How do we know with any confidence that our results will generalize to other occupations?

Although we cannot match in qualitative research the kind of sampling schemes and control mechanisms that quantitative research enjoys, we did design our study with the intent of making the broadest claims possible. For this reason, we did not choose a single firm within each of structural and hardware engineering but three firms within each. We did not choose a single group in our automotive engineering firm but three groups. We did not observe a single engineer at each firm but a cadre of engineers. We did not stay at each firm for a few days or weeks; we stayed for months. Moreover, we chose firms whose target clientele or final product varied so that we could cover a broad swath of the occupation. For example, in structural engineering, we chose one firm whose primary clientele were governments eager to seismically retrofit civic buildings (in a catastrophe, the public needs to know that the government is literally still standing), another firm whose clients were the developers of multistory commercial buildings (and whose concerns were primarily to make money, not ensure stability), and a third firm whose customers were semiconductor manufacturers desirous of one-story tilt-up fabrication plants (whose foremost objective was to quickly erect production capacity). We did, however, keep the generalized product the same: all three firms specialized in buildings, not bridges, highways arches, stadiums, or other structures. The similarity in generalized product helped us make useful comparisons across firms, while the variation among clients and type of building helped ensure that we considered the range of factors that might affect this occupation's technology choices.

Finally, we did not choose a single engineering occupation, or even just two; we chose three. Three occupations, in our case, ensured some variation across the manifestations of occupational factors (e.g., including both low and high task interdependence) as well as across the technology choices (e.g., hitting both the embrace and the shunning of automated links). That variation permitted comparisons that strengthened our analyses.

It is true that we did not stray outside engineering. We restricted ourselves to engineering occupations because we wanted, to the greatest extent possible, to control for work, technology, and product complexity. We wanted to control for work and technology so that we could directly examine task-technology fit arguments. We wanted to control for product complexity so that we could rule out what we thought would be the most obvious challenge to our claims, namely, that the designers of simpler products can make different technology choices than the designers of complex products need to make, independent of all other factors.

To argue that our restriction to engineering occupations somehow limits our claims about an occupational perspective would be to suggest that engineering is so fundamentally different from other disciplines that our findings must be specific to it. Is that true? Engineers create, as do bakers, artists, chefs, brewmasters, musicians, and carpenters, among others. Engineers are highly educated in a specialized field, as are doctors, lawyers, accountants, and nurses, to name a few; like these practitioners, they spend much of their working time trying to solve problems and tease out what is happening in a situation. Engineers are employees, just like marketers, sales staff, finance managers, and human resource trainers. In fact, we chose engineering in part because engineers are highly representative of the knowledge workers whose workforce ranks are swelling. We did not choose a niche discipline but one central to the economy.

All of this is not to say that having more occupations in our study would not have strengthened our findings. Having more occupations would very likely have helped us fine-tune our claims by, for example, allowing us to investigate more thoroughly the predictive ability of each occupational factor, and perhaps by bringing to light other occupational factors that did not materialize in our sample. But every researcher has to draw the line somewhere, and we think we drew the line around a fairly large area of empirical investigation. We welcome future studies of other occupations that might refine and extend our findings.

Method Too Onerous for Practical Use. The final limitation we consider is that, to understand technology choices, an occupational perspective requires in-depth knowledge of an occupation. In this sense, an occupational perspective is demanding of time and resources, and researchers can employ it only if they can gather the necessary information. In our study, we gleaned this information through our extended field observations, our interviews, and the time we spent studying and learning about each occupation. The process was long and tedious, as we fully describe in the appendix. In this respect, although an occupational perspective permits predictions and generalizations that voluntarist perspectives such as sociomateriality, critical realism, and social constructivism cannot, it surely cannot beat a technological determinist approach in terms of the effort required to generate these predictions and generalizations.

The potential of an occupational perspective, then, at its extreme, is simply this: a social constructivist perspective says that to know the technology choices of any setting, one must study that setting, which means that to know the choices of all settings, one must study them all. A technological determinist perspective says that to know the technology choices of any setting, one must study that setting or any other, which means that to know the choices of all settings, one need study only one. An occupational perspective says that to know the technology choices of any setting, one must study at least one setting of that occupation, which means that to know the choices of all settings, one must study at least one setting for each occupation.

A more muted interpretation is to say that an occupational perspective suggests we would do well to consider the host of occupational factors at play when investigating the technology choices in any given setting. Future scholars may have insight into methods that would allow them to ascertain and assess these factors without engaging in the kind of detailed fieldwork that we carried out. Absent that kind of insight, scholars face little choice, if they want to fully understand technology choices, than to take an empirical approach similar to our own. In the appendix that follows, we lay out our approach in detail and outline a few lessons learned that may lighten the load for future researchers.

Appendix: Research Design and Methods

We built the arguments in our book on rich qualitative data that our research team of twenty-seven individuals collected through field observations and interviews of engineers across three occupations, nine engineering groups, seven firms, and eight countries. In total, we logged more than one thousand hours of observation of more than one hundred engineers at work and wrote thousands of pages of field notes documenting what we observed. In addition, we brought more than a thousand work artifacts back with us from the field for further study. This appendix describes how we organized our research, trained our fieldworkers, and collected, managed, and analyzed our data. We also note a few lessons that we learned, in the hope that we might lessen the load for researchers thinking about carrying out similar inquiries.

Occupations, Firms, and Sites

With Steve Barley, her colleague at the time at Stanford University, Diane began a study of engineering work and technology with the intent of comparing occupations across two disciplines, civil engineering and electrical engineering. Steve and Diane thought that these two disciplines would provide a useful contrast because a century separated their emergence in the United States, with civil engineering arising in the 1770s and electrical engineering in the 1880s. Additionally, much of the knowledge in civil engineering, as we have noted elsewhere in the book, is enduring. For example, in the subfield of structural engineering, the load that a steel beam that is 18 inches wide and 35 pounds per linear foot can bear is much the same today as it was decades ago, and the principles that guide what makes buildings stand or fall are the basic, unchanging laws of physics, the understanding of which dates back to at least Newton. Knowledge in electrical engineering, by contrast, is renowned for how quickly it changes.

In the specialty of hardware engineering, programming languages, design rules, production techniques, and materials change every few years, if not months. Steve and Diane suspected that the difference in the rate of knowledge change between these two occupations might have important ramifications for technology choices among the engineers in them. For example, they surmised that technologies might become obsolete faster in hardware engineering than in structural engineering because the knowledge undergirding the technologies would become invalid or otherwise lose value.

Steve and Diane also appreciated that differences in occupational constraints, such as professional licensing, might play out in interesting ways, including how transparent solutions had to be and, therefore, the extent to which software applications might play a role in crafting solutions. At the time of the study's design, these two differences—rate of knowledge change and professional licensing—were the only occupational differences Steve and Diane were aware of. In other words, the occupational factors we have discussed in this book became known to our team almost entirely during the course of our research.

The addition of mechanical engineering in the form of automotive engineering to our study was as much happenstance as plan, which is not atypical for ethnographic work. IAC's R&D group had entered into an agreement with Stanford's Management Science & Engineering department to fund research, including our study of engineering work and technology. At the end of the first year of that agreement, IAC's R&D personnel approached Diane and asked her to include their engineers in the study. Diane was at first reluctant because IAC's engineering center in Michigan was halfway across the country from her home in California, a distance that posed significant problems for the kind of long-term field observations that had characterized the study to date. She thought that such intensive study was necessary to understand the nuances of engineering work in each occupation and to appreciate how engineers made technology choices. IAC offered to solve the problem by having a dozen of its R&D staff, most of whom held doctorates in such fields as statistics, sociology, and operations research, conduct the field observations under Diane's direction. Steve and Diane complemented these R&D staff with Stanford students, who did stints of observation over summer breaks. In the case of Paul, who was a graduate student at Stanford at the time, the stint extended over the course of a full academic year.

As it turned out, the addition of automotive engineers proved fortuitous for our work because these engineers occupied a midpoint between structural engineers and hardware engineers in terms of the role they allocated to the computer. In addition, this occupation was experimenting with using advanced information and communication technologies to facilitate the digital offshoring of engineering work around the globe, a choice that none of the firms in the other two occupations in our study had made. Thus, the addition of automotive engineers deepened and extended our study.

Our team examined three structural engineering firms and three hardware engineering firms to ensure that our findings were not particular to a given firm and to better determine whether similarities existed within occupations and differences across them. The six firms were located in the San Francisco Bay Area. We could not maintain this design for automotive engineering because it would have meant forming a sample of automotive firms nearly equal in size to the entire population of U.S. automotive firms (namely, the "Big Three"). Instead, we focused on IAC, and picked three engineering groups within that firm to study: Body Structures, Noise & Vibration, and Safety & Crashworthiness. Notably, these three engineering groups within IAC were larger than all three of the structural engineering firms that we studied and similar in size to one of the hardware engineering firms in our sample. We repeat in table 8.1 our table from the introduction that lists the firms and their market focus, and, in the case of automotive engineering, the groups we studied.

The three structural engineering firms in our study were professional firms in the sense that the only profession represented was structural engineering. Two of the firms each had a single office site with a staff of fifteen to twenty structural engineers. The third firm had four engineers at the site

Table 8.1
Firm Names and Market Focus

Occupation	Firm/Group Name	Market Focus
Automotive engineering	International Automobile Corporation (IAC) • Body Structures • Noise & Vibration • Safety & Crashworthiness	Cars and trucks
Structural engineering	Seismic Specialists	Civic building retrofitting
	Tall Steel	Multistory commercial projects
	Tilt-Up Designs	Semiconductor fabrication plants
Hardware engineering	AppCore	Microprocessor architectures and cores
	Configurable Solutions	Customizable microprocessors
	Programmable Devices	Programmable logic devices

we studied. That firm had two sister sites to the one we studied. One sister site, in Colorado, carried out independent projects with a group of six engineers and interacted very little with the site we studied; the other sister site, in Southern California, had two engineers who performed project management duties on the construction site of one of the major projects that we observed. We sent a researcher to visit each of these sister sites on a two-day trip to Colorado and a one-day trip to Southern California. We did these trips for the purpose of completing our understanding of the firm and gaining insight into senior engineers' extended work outside the office. Almost the entirety of our fieldwork with this firm, however, rested with the Bay Area office, and for this reason we count it as our only site for this firm. The remaining staff at each structural engineering firm was small and consisted, for the most part, of an office/payroll manager and a receptionist. As table 8.1 notes, one structural engineering firm specialized in seismic upgrades, particularly for civic buildings. The second firm specialized in single-story fabrication plants for the computer chip industry in Silicon Valley, and the third firm worked with developers and large clients on multistory commercial projects.

All of the hardware engineering firms in our study engaged in the marketing, design, and sale of products. One of the firms created microprocessor architectures and cores, another firm developed customizable microprocessor cores and peripherals, while the third firm designed programmable logic devices. All of the firms employed large numbers of software and hardware engineers in their design groups. Unlike the structural engineering firms, the hardware engineering firms also had sizable sales and marketing staffs, as well as individuals employed in such areas as finance, accounting, human resources, office management, and reception. Firm sizes ranged from two hundred to two thousand employees, with the number of hardware engineers ranging from thirty to more than a hundred.

At its peak in 2000, IAC employed 44,000 salaried workers, including more than 12,000 engineers. Its size was formidable compared to the other firms in our study. IAC had its headquarters in Michigan, with other engineering centers located in China, India, Korea, Australia, Mexico, Brazil, Germany, and Sweden. Within IAC's engineering division, the three groups that we studied—Body Structures, Noise & Vibration, and Safety & Crashworthiness—each had on the order of 130 engineers at the U.S. engineering center outside Detroit, Michigan, with more staff in each of these groups in each of the other global engineering locations. Each center except India built its own vehicle programs, which meant that each center except India had styling studios, test garages, proving grounds, design engineering groups, and analysis engineering groups, in addition to marketing, sales, finance, and the like. In India, as we have explained elsewhere, there were only digital engineering services, and no physical test facilities: IAC created the India site for the specific purpose of providing digital engineering services to the other sites worldwide. We visited every IAC site except the one in China, which at the time of our study appeared to be somewhat distinct from the rest of the network, making access difficult for us.

In total, we had seven firms in our sample: three firms each in structural engineering and hardware engineering and one firm in automotive engineering. Because the structural and hardware engineering firms did not subdivide their workforce into permanent groups (rather, temporary projects), we had one group to study at each of these firms and three groups at the automotive firm, for a total of nine groups. Finally, with the addition of the international sites at IAC, our fieldwork expanded to include eight countries: the United States, Mexico, Brazil, Australia, India, Korea, Sweden, and Germany. This international distribution meant that our sites, if we exclude the two Tall Steel sites that we visited only briefly, totaled fourteen: three sites each for structural and hardware engineering and eight sites for automotive engineering.[1]

Our Team

All told, over time, twenty-seven people worked on this research project. Among them were thirteen students, almost all of whom were from Stanford University, who collected data primarily by observing engineers at work, and occasionally by interviewing them. Fabrizio Ferraro, Julie Gainsburg, and Menahem Gefen observed structural engineers; Mahesh Bhatia, Jan Chong, Carlos Rodriguez-Lluesma, and Lesley Sept observed hardware engineers; and Vishal Arya, Will Barley, Daisy Chung, Aamir Farroq, and Alex Gurevich observed automotive engineers. Daisy Chung and Jeffrey Treem interviewed automotive engineers. Additionally, eleven R&D personnel from IAC collected data through field observations at the firm's Michigan headquarters. Jan Benson, John Caféo, Ching-Shan Cheng, Mike Johnson, Bill Jordan, Hallie Kintner, Mark Neale, Susan Owen, Dan Reaume, R. Jean Ruth, and Randy Urbance participated in this effort. Hallie Kintner spent a semester in residence at Stanford to work with us on early data analysis; she also interviewed engineers in Sweden.

Three university professors managed this project. Steve Barley from Stanford University worked with Diane to write the early grants for this project and to negotiate firms' participation in the study, a task that was particularly difficult among hardware engineering firms. Steve helped supervise data collection in automotive engineering, reading and commenting on students' observation notes to improve their skills in fieldwork. Steve's most instrumental contribution, however, was in helping conceptualize and write our findings for many of the study's scholarly articles, which to date number more than a dozen publications across top organization studies, communication, engineering, information studies, and education journals. His influence and the outcomes of his effort appear throughout this book.

While still a Stanford student, Paul Leonardi observed automotive engineers in Michigan over a period of twelve months, in India over two months, and in Mexico over four months. He also interviewed engineers and managers at each of these sites. As a professor at Northwestern University, Paul took the reins in our global data collection in automotive engineering, leading trips to Germany, Australia, India, Mexico, and Michigan. Paul also took the lead in much of our data analysis and in conceptualizing and writing a number of our journal papers. Moreover, he involved his own doctoral students in our research endeavors. He published a separate book based on his dissertation research, a field study of technology development, implementation, and use at IAC, on which we drew heavily in chapter 4.[2]

Diane Bailey directed this project, first from Stanford University and then from the University of Texas at Austin. She conducted over half of the observations of the engineers in structural and hardware engineering; she also interviewed engineers and managers in these two fields. In automotive engineering, Diane spent a month observing and interviewing engineers in India; she interviewed managers in Michigan; and she interviewed managers and engineers in Brazil and Korea. Diane developed the study's methods for data collection and analysis and trained all team members in these methods. She handled the bulk of the data management, conducted much of the data analysis, and led the conceptualization and writing of several of our journal publications.

Data Collection

We began our observations of engineers in June 1999 and ended them in July 2006. Table 8.2 shows when we observed at each site and how many observations we conducted. As the table indicates, we began with structural engineers, moved next to hardware engineers, and ended with automotive engineers. Within IAC, we observed engineers at only three of the eight sites in our study: Michigan, Mexico, and India. We did not observe engineers in the other countries because, in all of them except Australia, we did

Table 8.2
Schedule of Observations by Occupation and Firm/Location

Firm or Location	Observations Began	Observations Ended	No. of Observations
Structural Engineering (90 total observations)			
Seismic Specialists	June 1999	Sep. 1999	23
Tall Steel	Nov. 1999	June 2000	40
Tilt-Up Designs	Oct. 2000	Mar. 2001	27
Hardware Engineering (90 total observations)			
AppCore	Aug. 2000	Dec. 2000	25
Configurable Solutions	Feb. 2001	July 2001	40
Programmable Devices	Feb. 2002	Apr. 2002	25
Automotive Engineering (IAC, 192 total observations)			
Michigan	July 2003	Nov. 2006	156
Mexico	Jan. 2006	Apr. 2006	11
India	Apr. 2006	July 2006	25

not speak the language, which would have made observation of everyday work impossible. The larger reason, though, is that by the time we had observed engineers at these three sites, we were comfortable enough in our knowledge of work practices and technology choices to gain the rest of what we needed to know through other means. Thus, we complemented our observations of automotive engineers at IAC with interviews, surveys, and archival records to help us understand how the firm organized the work that it distributed globally via technology. Table 8.3 details that data collection and its timing. In total, our data collection spanned more than a decade, from 1999 to 2010.

As the tables indicate, we carried out our observations and interviews sequentially across the three engineering occupations, a process that allowed us to come to grips with the challenges of this fieldwork. Engineering work has been notoriously difficult for scholars of work to study.[3] Although Diane holds three degrees in engineering, her education was in industrial engineering, not structural, hardware, or automotive engineering. Thus, whereas she understood universal engineering paradigms of problem solving and model building, she was unfamiliar with the analyses, terminology, and technologies of these occupations. The rest of our team was no better prepared initially for this endeavor. Simply being a skilled observer is likely to be insufficient for documenting what engineers do all day. To understand why engineers make the technology choices that they

Table 8.3
Interviews at IAC by Year and Site

Year	Interviews	
	Site	*n*
2006	India	18
	United States	25
	Total	*43*
2008	India	45
	United States	54
	Australia	17
	Brazil	8
	Germany	36
	Korea	15
	Mexico	28
	Sweden	11
	Total	*204*
2010	India	35
	United States	31
	Total	*66*
Total		**313**

do at work and why they act as they do, observers need scientific and technical knowledge that many scholars of work lack. To make matters worse, engineers speak in technical tongues that are specific to their specialty, their tasks have few everyday correlates, they render mathematical calculations quickly, and they make use of sophisticated tools whose functioning is not easily gleaned. The number of people whom engineers consult and the number of technologies they use also complicate observation. At any point in time, an engineer may employ simultaneously three to four technologies, engage other engineers in discussion, or consult a small library of documents and electronic files. As a result, observers must work hard to hear what is said and to notice what is done.

We took steps to meet these challenges. Before entering the field, our research team read texts and course notes from relevant civil, electrical, and mechanical engineering classes. Professional engineers (not in our study) tutored us on design tasks, analysis tasks, products, and technologies that we would encounter. Initially, in structural engineering, two researchers jointly shadowed the engineers. Working as a pair, one researcher recorded events, while the other documented only technical terms and their meaning. From the latter's field notes we developed technical glossaries to help us better hear and understand what was later said in solo observations. We took this approach because our earliest field notes from the sites missed more words in dialogues than they captured. We also bought a dictionary of civil engineering terms. By the time we had learned about fifty or so critical everyday terms, we were able to revert to having a single researcher on each observation. Examples of structural engineering terms that we had to learn include TS2×2, W18×35, RFI, details, kips, Unistrut, moment frame, ICBO numbers, pile caps, caissons, unity check, and analyses of three types (gravity, lateral, and foundation). As fieldwork proceeded, we purchased and studied other books and manuals that we observed engineers using, including manuals on steel design, structural engineering textbooks, code manuals, and vendor catalogs. Doing so helped us to understand what kinds of information the engineers sought in these resources and provided us with greater understanding of the problem-solving and model-building paradigms in this field. We also tried to read past observation notes of an informant before conducting a new observation of that person to prepare us for the technologies and projects we would be likely to encounter in the field that day, and to remind us of areas of confusion that we might clear up with a question at the beginning or end of an observation.

These efforts on our part to learn the fundamentals of engineering design and analysis through reading, vocabulary building, studying, and lessons were greatest in structural engineering. In hardware engineering and automotive engineering, we repeated our requests for lessons in the design and analysis process and in primary technologies, and we purchased a handful of textbooks and manuals from these fields to accelerate our learning and to provide definitions of foreign terminology. We did not, however, need to double up on observers to capture words we might otherwise have missed when observing hardware or automotive engineers. Perhaps we were better able to capture terminology in these fields as compared to structural engineering because in hardware engineering so many of the terms appeared in written form in code on a screen, where we could see them (e.g., BUFG mux, SysIOPAddr), and in automotive engineering many of the terms were familiar to us as car parts. These two fields also had more of an online presence at that time than did structural engineering, as evident, for example, in the technology forums in which hardware engineers engaged, thus providing us with further resources for reading up on terms we did not know. In addition, hardware engineering and automotive engineering had more of a presence in the popular and professional media than did structural engineering, so that we could complement our understanding of these occupations with background material from popular books, national newspapers, and business-oriented magazines.

Initially, for each occupation, we attempted to verify our understanding of what we had seen and heard at the end of an observation session by asking engineers to recap what had transpired. We quickly found, however, that engineers recapped a day's events by employing the same language we were already struggling to understand or leaving out the same information we had missed in real time. To avoid this problem, we began to recap what we thought had transpired in our own terms. This practice allowed engineers to assess our level of understanding and provide background information when needed. As we became more conversant with the engineers' tasks, technologies, and language, we were able to revert to having the engineers recap events, and eventually we required no recaps whatsoever.

In addition to modifying our approach to collecting data, we also had to develop special techniques for recording an engineer's actions and words because our study demanded that we not overlook any technology use. Fortunately, we were more observer than participant in the field (no one, it turned out, entrusted us as aides in the design or analysis of automobiles, buildings, or chips), which meant we had no duties other than to collect data for our study. Therefore, unlike traditional ethnographers, whose time in the field might be consumed with the carrying out of rituals, tasks, jobs, and roles in, for example, the street, a gang, or a factory (akin to how novices engage in legitimate peripheral participation on first entering a community of practice) so that they might, as ethnographers, be accepted into the scene, and who as a result might have time only for hurried snatches of notes on conversations and jottings of events to serve as jolts to the memory at the end of the day, we could devote ourselves during our observations to the task of building detailed records of engineers' action and speech.[4] No one in the field mistook us as engineers-in-training; everyone understood we were there to take careful notes on what they did and how they did it for the sole purpose of thinking about technology choices, and no one was surprised to see us occupied fervently with pen, pad, and recorder, although they were intrigued and at times amused by our interest in what they often considered the mundane. That our purpose was so well understood and accepted by our informants is reflected in how many times they greeted an observer by saying, "Today is going to be boring for you. I'm hardly using any technology."[5]

Our special recording techniques began with noting the passage of time in our field notes in ten-minute intervals. Between these time stamps, we took running notes on engineers' every interaction with technology, people, documents, and other artifacts, and we audiotaped conversations when we could not keep pace by writing notes or when discussions became very technical. When our informants worked at computers, we requested screenshots of software interfaces and digital copies of the models, code, email, and documents they worked on or with during the session. Each day we photocopied key documents the engineers had employed, such as drawings, sketches, pages from brochures, handwritten calculations, and scraps of paper on which they had scribbled notes. We found that attending to the details of engineering work in this manner made it difficult for us to observe an engineer longer than three or four hours a day; by then, our hands were cramped, our eyes strained, and our minds unable to maintain intense focus.

Although this work was exhausting, collecting so much data and so many types of data in the field paid huge dividends. Our data collection methods enabled us to prepare extensive field notes that described actions, conversations, and visual images simultaneously, and therefore to produce a record not only of what engineers did and said but also of what they worked on and created. The first step in assembling a day's field notes was to expand the running notes taken in the field into full narratives that someone who had not been on site could understand. This step was particularly important insofar as we had a large team of researchers, many of whom, as students, were transient, in that their stay with the project lasted only a year or two. We needed to make sure that the rest of us on the team could read the notes of those who would leave long after they had departed.

With this narrative in hand, we transcribed tapes directly into the body of the text at the point where the talk occurred. Similarly, we indexed screenshots and photocopies of documents at the point in the field notes where they were used. Weaving together actions, conversations, and images allowed us to capture and better understand an engineer's gestures and movements, especially the referents of indexical speech (e.g., "Here you see ..." or "Right here we need ..."). We marked up drawings, for example, by noting during a design session which section attracted the engineers' attention at which point in the conversation, so that a progression of markings on the drawing corresponded to the conversation's path. We wrote appendices for each day's field notes that described the artifacts we collected and, if the artifact was the result of the engineer's work, how it evolved. A description of a spreadsheet, for example, might note which numbers in the cells existed at the time the engineer opened the file, which numbers he added or deleted, and which numbers were automatically calculated by embedded formulas. These descriptions also contained information about how and why the engineer employed the document, and who had created it. Finally, we consulted the web for information on commercially available applications that the engineers employed and included this information in appendices. To reduce the odds of forgetting, we began expanding our field notes immediately after leaving the site. As compared to most ethnographers, however, who complete the writing up of a day's field notes in a single evening, we required two to two and a half days to complete a full narrative of a half-day's observation.

Consistency and coordination were especially important in a project spanning so many researchers, years, and sites. Mandating that all field notes conform to a standard format contributed to consistency across researchers. These standards included rules for noting action within dialogue (square brackets to set action off from words), for indexing and describing documents (e.g., noting what software application produced or displayed the document), and for listing at the beginning of a day's notes the major technologies the engineers used and tasks the engineers performed. Diane trained all team members who did fieldwork and reviewed every set of notes they produced for thoroughness, technical accuracy, and conformity to formats. For IAC's R&D staff, this training consisted of three days of on-site instruction and practice in Michigan, followed by commenting on and rewriting of field notes for a period of several weeks. For students doing summer observations at IAC, this training involved a week of on-site instruction and practice at Stanford, followed similarly by commenting on and rewriting of field notes over several subsequent weeks. (Steve and Diane shared this work.) At the end of the summer, the students had a one-week debriefing session back at Stanford to ensure their notes were complete and conformed to all standards. For doctoral students working on the project throughout the year, the training was more gradual, but involved the same elements of instruction, practice, commenting, and rewriting. We held research team meetings as necessary to devise solutions to difficulties encountered in the field, for example, to arrange tutoring sessions on a new technology or to acquire a manual that the team needed to read.

Unlike our observations, our interviews followed the norms of ethnographic techniques. We transcribed all our interviews; in the early years we did this work ourselves, but later we hired professional transcriptionists when the volume of interviews grew. Interviews ranged from thirty to ninety minutes in length. Interviews with engineers tended to be shorter than those with managers, largely because we often had time to informally interview engineers during lulls in observations. We created protocols for our formal but semistructured interviews of managers and engineers in our

2006–2010 fieldwork at IAC, the focus of which was the global distribution of engineering work. We used these protocols as guides in our interviews, but we felt free to stray from the protocols when aspects of the particular context suggested different avenues of exploration or when the informant was keen to discuss a particular event, state of affairs, or viewpoint. Examples of questions in our protocol for engineers sending digitized work to IAC's India site from their other technical centers around the world included the following:

- Almost every engineer has a lot on his plate, so you need to decide which work to do yourself and which to send to India. What guides that decision for you?
- Can you explain the process you go through in preparing projects to send to India?
- How do engineers in India return work to you? (Do they simply send you the files?)
- How much visibility do you have into the way projects are organized and work is done at [IAC's India location]?
- Do you talk to engineers in India more or less than you talk to engineers at other sites?
- What type of guidance do you get from others about working with people in India?

Examples of questions for engineers at the India site who performed digital engineering services for engineers at IAC's other global centers included

- How clear are the instructions or request you receive [from engineers worldwide]?
- What kinds of things do you need to have clarified in the instructions or requests you receive?
- What is the server situation like? (Must you wait for space or licenses to run jobs?)
- What do you view as your expertise?
- What kinds of things typically cause return of assignments or requests for rework?
- What kinds of misunderstandings happen?
- How would you characterize your relationship with the engineers who send you work requests from abroad?

For our informal interviews, which came up in the course of our observational fieldwork, we had no set protocol, preferring instead to ask the questions that were necessary to further our understanding in each occupation.

For example, in structural engineering, we asked Sheila at Seismic Specialists, "Can you tell me what would drive the choice between Beams2 and WFB [two analytical computer programs]?" and "One program you haven't mentioned yet that you used at all is Biax. What does Biax do and why haven't you mentioned it?" In hardware engineering, we asked Randy at Configurable Solutions, "Well, didn't you have programming classes in college where you finally learned how to write good code? Or that just didn't happen?" and "Do you really know who is in charge of a part [component]? Is it clear?" In general, these questions were particular to the engineer and to our history of observation with the engineer, and for this reason we typically did not repeat them in other contexts.

We maintained digital files of all our observation and interview notes. We also maintained one complete set of hard-copy notes in a set of thirty-three two-inch binders; this set included the originals of all documents and other artifacts that we took from the field. From the digital files we created databases of notes for use in Atlas.ti, the qualitative analysis software application that we employed in our analyses. We also created an Access database for scans of all the documents and other artifacts that we took from the field.

Analyzing Data

We employed a variety of analytical techniques to help us interrogate our data around technology choices. Although these techniques differed depending on which technology choice we were exploring, undergirding each of them was the close and repeated reading of our notes.[6] Without that reading and the understanding it generated, we could not have fathomed how to begin our analyses. As we noted in our description of data collection, it took us two and a half days to prepare the notes for a single observation; that time served as our first close reading of the notes. Steve and Diane read many of the team's notes in the course of training individuals to work in the field. Before each visit to a site to observe an engineer, we tried to read the completed field notes from prior observations of that informant, a practice that served as yet another reading of the notes. Before ending our team's stay at any site and concluding our fieldwork there, we drew up tables of tasks and technologies, which we reviewed with our informants to ensure that we had seen all the major activities that constituted their work and that we had not missed the use of important technologies. The creation of these tables required us to read all the field notes from that site so that we might inventory, as it were, tasks and technologies, thus providing another opportunity for close reading. Finally, before undertaking each analysis, we immersed ourselves in the notes with the purpose of focusing our attention on those aspects of the notes most pertinent to the current analysis. Overall, we made close reading of the notes a priority throughout the process of collecting and analyzing data.

The first and most immediate outcome of our close reading of the notes was our realization that engineers within occupations made similar technology choices, and engineers across occupations either made different technology choices or else made the same choices but for different reasons. This realization caused us, while still collecting data, to pay attention to what occupational factors may have shaped engineers' choices, and to probe this question in casual conversations with engineers during observations as well as interviews with engineers and managers separate from the observations. Afterward, following the advice of case study expert Robert Yin,[7] we organized our analysis around this motivation, namely, the identification of which occupational factors, if any, shaped engineers' technology choices.

Drawing on the recommendations of experts in qualitative data analysis,[8] we began with a within-case analysis in which we examined each relationship (human-technology, technology-technology, and human-human) and occupation (structural, hardware, and automotive engineering) combination in turn. We built separate narratives for each of these relationship-occupation combinations. In the case of the human-technology relationship, for example, we began with structural engineering and documented, through our inventory of structural engineering tasks and technologies, in conjunction with our field notes, what tasks engineers allocated to the computer and what tasks they did not. Then we turned to our observation and interview data to find evidence of factors that shaped the choices that led to that set of allocations. To help us think through the influence of each factor on an occupation's choice, we asked ourselves thought questions, such as, "What if this factor had been high instead of low in this occupation? Would that have made a difference?" We repeated this analysis with each relationship-occupation combination until we had the basic narratives for chapters 3, 4, and 5.

In the course of building the narratives for chapters 4 and 5, we drew on some of our previously published work that had appeared as articles in scholarly journals. We refrain from repeating here all the details that appear in the methods sections of those articles. We do, however, offer basic outlines of what kinds of analyses we performed, and provide citations to those works.

Specifically, in chapter 4, in addition to drawing on Paul's published case study of IAC, we turned to our prior article with Jan Chong, a Stanford doctoral student at the time, on technology interdependence to introduce the concepts of technology gap, gap width, and workflow direction.[9] The first task of data analysis in that article was to conceptualize technology interdependence in a manner that would allow us to identify and isolate episodes in the field notes in which interdependence was evidenced. We adopted the recommendation of Barney Glaser,[10] an expert in qualitative research, to develop analytical constructs, grounded in our observational data, that would aid us in systematic analysis, and we followed his practice of "theoretical sensitivity" to develop these concepts. Ultimately, we identified 310 gap encounters, or episodes in which a technology gap appeared in the course of action, in our field notes of structural engineers and hardware engineers. Diane coded each of these 310 gap encounters using Atlas.ti to determine gap width and workflow direction, which we discussed in chapter 4.

In chapter 5, we drew on Diane and Steve's article on teaching and learning ecologies to help build the narrative for structural engineers' choice to reject distant work.[11] The first task of data analysis in that article was to separate learning episodes from the stream of behavior recorded in our field notes. For that we turned to social psychologist Roger Barker's work for help.[12] Although Barker and his students were influenced by anthropology, their approach to analyzing data from observations of people differed significantly from how most ethnographers treat field notes.[13] Whereas ethnographers typically look for meaning and themes, Barker and his associates parsed their field notes into "behavioral episodes," distinct units of analysis whose frequency they could count and whose distribution they could analyze, thus allowing comparisons across settings. The value of Barker's approach is that it allows one to document actions precisely over time and space, which is difficult to achieve when analysis targets meanings and themes. That approach was exactly what we needed to help us tease out learning episodes in the engineering workplace, and our detailed field notes readily lent themselves to the kind of analysis that Barker favored. Full details of our method appear in our journal article.

Teaching and learning had little relevance in explaining automotive engineers' choice to employ distant work. Thus, when discussing automotive engineers in chapter 5, we drew on Diane, Paul, and Steve's article examining the role of physical and digital artifacts in this work.[14] In particular, we focused on how the increasing verisimilitude of crash simulations led engineering managers to believe they could offshore engineering analysis tasks. Paul conducted the analysis of our data in this realm, following a case study approach similar to what we employed for our analysis of factors but with a stronger historical bent. His historical analysis began with the predigital period of engineering analysis and continued through the arrival of CAD technologies, advanced simulation technologies, and finally the telecommunications technologies that made possible the electronic transport of digital work artifacts.

Having completed the narratives for each relationship-occupation combination, we turned to cross-case analysis, which became the foundation for chapter 6. We began by staying within occupation and looking across relationships to ask why a particular factor affected one choice for that occupation and not another. For example, in hardware engineering we asked why product interdependence played a role in the human-human relationship but not the other two relationships. We also asked ourselves why certain factors played no strong role in any relationship (e.g., technology cost in hardware engineering, despite the high cost of technology in this field), and why other factors played a strong role in each of them (e.g., competition in hardware engineering).

We next held relationship constant and looked across occupations to investigate questions such as why one factor played a strong role in a given relationship in one occupation but not in the other occupations. This comparison forced us to consider in the context of one occupation factors that arose only in the others. Such comparisons caused us to return to our data to confirm that we had not somehow overlooked a factor, and we probed to understand why a particular factor may have played no role in a given occupation. In this manner, we were able to use what ethnographer and sociologist Jack Katz has called "negative observations" to help support our positive claims[15]: the lack of evidence in our field notes and interview transcripts for a factor's strong role lent credence to our claims about the influence of other factors. In other words, given our intensive data collection and our extended time in the field with each occupation, if a given factor had played a role in a technology choice, then we should have seen its impact in the same way that we saw the impact of other factors in that relationship-occupation combination, or in the same way that the factor showed its hand in another occupation for the same relationship.

In the course of these cross-case analyses, first within occupations but across relationships and then within relationships but across occupations, we conducted additional close readings of our field notes and interview transcripts, refining our narratives in the face of supporting or disconfirming evidence for our developing ideas. Combined, our within-case and cross-case analyses helped us understand the relative importance of each factor within occupations and relationships.

All the analyses rested on our initial assessment of the manifestation of a factor within an occupation (e.g., that liability was high in structural and automotive engineering but low in hardware engineering). We determined these manifestations through our field notes, interview transcripts, and archival work as necessary. For example, we knew that competition was bid-based in structural engineering through our early interviews with the presidents of these firms; we knew competition was time-based in hardware engineering and automotive engineering through our early interviews with senior managers in these fields and through what we read about the occupations in popular books, national newspapers, and business-oriented magazines. These early indications were confirmed in our fieldwork through comments that our informants made, some of which we quoted in this book. That technology was not transparent in structural engineering, on the other hand, was not information we gained through interviews or archival sources; rather, we reached that determination through close reading of our field notes; similarly, we turned to our observation data to assess technology transparency in the other two occupations. Thus, manifestations of the factors revealed themselves through different means, which we corroborated via triangulation of our data as much as possible.[16]

We organized our results in chapters 3, 4, and 5 according to the relationship rather than the occupation because we wanted to place the emphasis on differences across occupations, a goal we could achieve more easily if the narratives for each occupation resided close to one another. In chapter 6 we drew across the narratives in chapters 3, 4, and 5 to build whole narratives for each occupation. We summarized the components of the narratives in tables 6.2 to 6.4. Then, to place our focus squarely on the factors themselves, we created table 6.5, which required no new analysis, simply the reorganization of results gained from our previous analyses.

The information that table 6.5 conveyed allowed us to tentatively explore the predictive potential of each factor, a step that constituted our final analytical act. We conducted that analysis simply by examining the choice-instantiation combinations for each factor. Because each choice in our study had only two alternatives (e.g., shun or embrace automated links in the case of the technology-technology relationship) and because most factors had only two instantiations (e.g., custom or mass production in the case of the market factor), our task was most often just to examine what patterns emerged in each two-by-two cell that represented a choice-instantiation combination. From that examination, we could derive tentative expectations of each factor's predictive potential.

Lessons Learned and Shared

Studies of the magnitude of our study in terms of time, resources, team, and data come with ample lessons learned. To aid scholars who may be considering quests similar to ours, we offer three lessons from our experience, in the hope that future scholars may avoid, or at least be prepared for, some of the pitfalls we encountered.

The first lesson we learned was that having a large team of people conducting observations enabled us to speed up our fieldwork considerably.[17] Each observation consumed three full days of a researcher's time: half a day to observe, and the remaining time to construct the complete record of field notes. That meant that at best, each researcher could do no more than two observations per week, a pace that few of us could maintain for more than a month or so before falling behind in documenting and maintaining our data properly. Only by having multiple observers could we collect data at a decent pace.

Although having a team was a lifesaver for us in terms of data collection, the upshot of all that effort was that we amassed a lot of data, much of which was in the form of written field notes, which Steve or Diane had to read and comment on so that the observers might make corrections. When we had four students in the field for the summer, for example, we had eight sets of field notes to review a week, each amounting to about twenty single-spaced pages, with attachments for artifacts. Reading, commenting on, tracking, and storing so much data were time-consuming activities. In retrospect, we should have hired a project manager, perhaps a postdoctoral student, to handle that aspect of the work for us. One tends not to think of research employing ethnographic techniques as needing a team, let alone a project manager, but our experience suggests that both have potential for large-scale projects.

The second lesson we learned is that engineering work was indeed challenging to study, particularly in the beginning. We struggled to understand what we were seeing in the field, and we struggled even more to understand what it meant theoretically back in our offices. We did not grasp the significance of what we had seen in the structural engineering workplaces until we entered the hardware engineering workplaces and began to register differences. Still, it took us months in the field to feel comfortable in our understanding of hardware engineering work. That meant that we had spent at least a year and a half in the field before we sensed the story that we might tell, and fully three years in the field before we had completed observations in the first two occupations. If we add in the time that coding and analyzing the data took, plus the time for writing up our work and negotiating the review process, quite a few years passed before the first publications from our research emerged. Although the payoffs in terms of depth and range of analysis that this large undertaking made possible were considerable, the risks, in particular a slow rate of publishing at the beginning of the project, were equally significant.

In our case, there was, unfortunately, no good way to speed up this process: Engineering work requires time to learn its contours. We suspect that other, less technical types of work may prove easier to study. For example, since collecting these data, we have been involved in similar studies of librarians, marketing professionals, offshore graphic designers, and remote bankers, and we have found that studying their work has required far less preparation on our part, with far fewer difficulties in understanding terminology, tasks, and technology, than in the case of the engineers. These experiences suggest to us that the time required studying technology choices may vary considerably by occupation.

The third lesson we learned was that our methods were in some cases novel, lying well outside the realm of traditional ethnography and, for that matter, most qualitative research. This novelty was evident, for example, in our analysis of teaching and learning episodes among structural and hardware engineers. Because our methods were novel, they required thorough explanation and justification beyond what one normally expects of ethnographers. Scholars in the knowledge literature, for example, were skeptical at first that we could capture episodes of learning opportunities in everyday action in the engineering workplace. That is why we turned to the work of social psychologist Roger Barker to provide evidence that at least one other scholar before us had similarly managed to bracket off episodes from ongoing everyday action. In other words, given the novelty of our methods and our claims, we needed to establish that what we did was possible. Our incorporation of artifacts and our detailed notes about them, their use, and their creation provided another example of how our methods differed from traditional ethnography, as did our detailed field notes, which wove together actions, speech, and objects. Only lengthy explanations of our data collection and analysis methods yielded the transparency that reviewers required and laid bare our claims to rigor, validity, and mere possibility.

In summary, our data collection methods lay somewhere between the precise attention to movement and detail of the industrial engineer and the jottings-converted-to-field notes approach of many traditional participant-observers who conduct ethnographies. Meanwhile, our data analysis methods extended beyond the ethnographer's traditional quest for meanings and themes, at times employing counts of types of events to establish patterns of behavior. Because novelty demands explanation and legitimation, we often had to write longer, more detailed methods sections than one might normally see in academic writing.

Notes

Introduction

1. Like all personal names in this book, the firm names are pseudonyms.

Chapter 1

1. Paul M. Leonardi and Stephen R. Barley. "What's Under Construction Here? Social Action, Materiality, and Power in Constructivist Studies of Technology and Organizing," *Academy of Management Annals* 4 (2010): 1–51.

2. Harold J. Leavitt and Thomas L. Whisler, "Management in the 1980s," *Harvard Business Review* 36, no. 6 (1958): 41–48.

3. Joan Woodward, *Management and Technology* (London: HMSO, 1958), 16.

4. Charles Perrow, "A Framework for the Comparative Analysis of Organizations," *American Sociological Review* 32 (1967): 194–208, at 195.

5. See, for example, I. R. Hoos, "When the Computer Takes over the Office," *Harvard Business Review* 38, no. 4 (1960): 102–112, and Thomas L. Whisler, *The Impact of Computers on Organizations* (New York: Praeger, 1970).

6. Hak Chong Lee, *The Impact of Electronic Data Processing upon the Patterns of Business Organizations and Administration* (Albany: State University of New York Press, 1965).

7. Jeffrey Pfeffer and Huseyin Leblebici, "Information Technology and Organizational Structure," *Pacific Sociological Review* 20, no. 2 (1977): 241–261.

8. See, for example, Peter M. Blau, Cecilia McHugh Falbe, William McKinley, and Phelps K. Tracy, "Technology and Organization in Manufacturing," *Administrative Science Quarterly* 21, no. 1 (1976): 20–40, and S. R. Klatzky, "Automation, Size, and the Locus of Decision Making: The Cascade Effect," *Journal of Business* 43, no. 2 (1970): 141–151.

9. Langdon Winner, "Prophets of Inevitability," *MIT Technology Review* 101, no. 2 (1998): 62.

10. Frank Levy and Richard J. Murnane, *The New Division of Labor: How Computers Are Creating the Next Job Market* (New York: Russell Sage Foundation, 2004).

11. Paul M. Fitts, ed., *Human Engineering for an Effective Air Navigation and Traffic Control System* (Washington, DC: National Research Council, 1951).

12. See Alphonse Chapanis, "On the Allocation of Functions between Men and Machines," *Occupational Psychology* 39, no. 1 (1965): 1–11.

13. For these details and others, see David A. Hounshell, *From the American System to Mass Production, 1800–1932* (Baltimore, MD: Johns Hopkins University Press, 1984), chap. 6, "The Ford Motor Company & the Rise of Mass Production in America," 217–261. See also David E. Nye, *America's Assembly Line* (Cambridge, MA: MIT Press, 2013), chap. 2, "Invention," 13–38.

14. Charles R. Walker and Robert H. Guest, *The Man on the Assembly Line* (Cambridge, MA: Harvard University Press, 1952).

15. Ely Chinoy, *Automobile Workers and the American Dream* (Garden City, NY: Doubleday, 1955), chap. 1, "Tradition and Reality," 1–11.

16. James P. Womack, Daniel T. Jones, and Daniel Roos, *The Machine That Changed the World* (New York: HarperCollins, 1990).

17. Paul S. Adler, "Time-and-Motion Regained," *Harvard Business Review* 71, no. 1 (1993): 97–108.

18. Critics of Japanese production techniques often claimed that plants like Saturn's Spring Hill facility were superior alternative implementations of assembly lines. These critics argued that Japanese methods required workers to work at accelerated paces while managers ignored the severe health problems (including stress, headaches, stomachaches, exhaustion, and carpal tunnel syndrome) that higher pace induced. In contrast, plants like the one at Spring Hill were said to promote healthy, safe, and socially rewarding work. These differences in opinion are not critical for our argument because the promoters of Japanese techniques as well as their critics challenged technologically deterministic accounts: both factions held that the manner in which the assembly line was implemented uniquely shaped work and worker outcomes. See Christian Berggren, *Alternatives to Lean Production* (Ithaca, NY: ILR Press, 1992); Laurie Graham, *On the Line at Subaru-Isuzu: The Japanese Model and the American Worker* (Ithaca, NY: ILR Press, 1995); Kim Moody, *Workers in a Lean World: Unions in the International Economy* (London: Verso, 1997); and James W. Rinehart, Christopher Victor Huxley, and David Robertson, *Just Another Car Factory? Lean Production and Its Discontents* (Ithaca, NY: ILR Press, 1997), among others. For an excellent overview of industry practices with respect to lean production and its

alternatives, see Thomas A. Kochan, Russell D. Lansbury, and John Paul MacDuffie, eds., *After Lean Production: Evolving Employment Practices in the World Auto Industry* (Ithaca, NY: Cornell University Press, 1995).

19. Harley Shaiken, Steve Lopez, and Isaac Mankita, "Two Routes to Team Production: Saturn and Chrysler Compared," *Industrial Relations* 36, no.1 (1997): 17–45.

20. Eileen Appelbaum and Peter Albin, "Computer Rationalization and the Transformation of Work: Lessons from the Insurance Industry," in *The Transformation of Work? Skill, Flexibility and the Labour Process*, ed. Stephen Wood (London: Unwin Hyman, 1989), 246–265.

21. Shoshana Zuboff, *In the Age of the Smart Machine: The Future of Work and Power* (New York: Basic Books, 1988).

22. Stephen R. Barley, "Technology as an Occasion for Structuring: Evidence from Observations of CT Scanners and the Social Order of Radiology Departments," *Administrative Science Quarterly* 31, no. 1 (1986): 78–108. See also Anthony Giddens, *Central Problems in Social Theory* (Berkeley: University of California Press, 1979), and idem, *The Constitution of Society: Outline of a Theory of Structuration* (Cambridge: Polity Press, 1984).

23. Wanda J. Orlikowski, "The Duality of Technology: Rethinking the Concept of Technology in Organizations," *Organization Science* 3, no. 3 (1992): 398–427.

24. Wanda J. Orlikowski, "Using Technology and Constituting Structures: A Practice Lens for Studying Technology in Organizations," *Organization Science* 11, no. 4 (2000): 404–428.

25. Ibid., 407.

26. Marie-Claude Boudreau and Daniel Robey, "Enacting Integrated Information Technology: A Human Agency Perspective," *Organization Science* 16, no. 1 (2005): 3–18.

27. See Michele H. Jackson, Marshall Scott Poole, and Tim Kuhn, "The Social Construction of Technology in Studies of the Workplace," in *Handbook of New Media*, ed. Leah A. Lievrouw and Sonia Livingstone (London: Sage, 2002), 236–253; and Jannis Kallinikos, *The Consequences of Information: Institutional Implications of Technological Change* (Cheltenham, UK: Edward Elgar, 2006).

28. Leonardi and Barley, "What's under Construction Here?," 33.

29. Our two-by-two matrix cloaks the possibility that the members of each pair may represent the ends of a continuum. We maintain the matrix for its simplicity and clarity in pointing out fundamental differences among approaches, and because trying to determine where each approach might sit on a continuum lies beyond our scope.

30. See, for example, Bruno Latour and Steve Woolgar, *Laboratory Life: The Construction of Scientific Facts* (Beverly Hills, CA: Sage, 1986).

31. See, for example, Karin D. Knorr-Cetina and Michael Mulkay, eds., *Emerging Principles in Social Studies of Science (*London: Sage, 1983).

32. Andrew Pickering, *The Mangle of Practice: Time, Agency, and Science* (Chicago: University of Chicago Press, 1995).

33. Joan H. Fujimura, "Sex Genes: A Critical Sociomaterial Approach to the Politics and Molecular Genetics of Sex Determination," *Signs: Journal of Women in Culture and Society* 32, no. 1 (2006): 49–82; Ian Hutchby, "Technologies, Texts and Affordances," *Sociology* 35, no. 2 (2001): 441–456.

34. Wanda J. Orlikowski, "Sociomaterial Practices: Exploring Technology at Work," *Organization Studies* 28, no. 9 (2007): 1435–1448, at 1437.

35. For recent examples of this work, see the following edited collection: Paul M. Leonardi, Bonnie A. Nardi, and Jannis Kallinikos, eds., *Materiality and Organizing: Social Interaction in a Technological World* (Oxford: Oxford University Press, 2012).

36. Alistair Mutch, "Actors and Networks or Agents and Structures: Towards a Realist View of Information Systems" *Organization* 9, no. 3 (2001): 477–496; Alistair Mutch,"Technology, Organization, and Structure: A Morphogenetic Approach," *Organization Science* 21, no. 2 (2010): 507–520; Olga Volkoff, Diane M. Strong, and Michael B. Elmes, "Technological Embeddedness and Organizational Change," *Organization Science* 18, no. 5 (2007): 832–848; and Olga Volkoff, and Diane M. Strong, "Critical Realism and Affordances: Theorizing IT-Associated Organizational Change Processes," *MIS Quarterly* 37, no. 3 (2013): 819–834.

37. See Roy Bhaskar, *The Possibility of Naturalism* (Hemel Hempstead, UK: Harvester, 1979).

38. Denis C. Phillips, *Philosophy, Science, and Social Inquiry* (Oxford: Pergamon, 1987), at 205.

39. For a discussion of how critical realism is related to sociomateriality, see Paul M. Leonardi, "Theoretical Foundations for the Study of Sociomateriality," *Information and Organization* 23, no. 2 (2013): 59–76.

40. For an example of one of the few empirical studies of critical realism in technology choices, see Volkoff, Strong, and Elmes, "Technological Embeddedness and Organizational Change," 2007. In a field study of the implementation of a new ERP system in a large industrial tools manufacturing firm, the authors showed that changes that emerged as the results of the technology's use were not gradual and emergent. Instead, changes came in fairly predictable waves as old routines met new features of the technology and as people had to make decisions about how to change routines.

41. For a summary of deskilling arguments, see Stanley Aronowitz and William DiFazio, *The Jobless Future: Sci-Tech and the Dogma of Work* (Minneapolis: University of Minnesota Press, 1994), chap. 3, "The End of Skill?," 81–103.

42. Barbara Garson, *The Electronic Sweatshop: How Computers Are Transforming the Office of the Future into the Factory of the Past* (New York: Simon & Schuster, 1988).

43. Harley Shaiken, *Work Transformed: Automation and Labor in the Computer Age* (New York: Holt, Rinehart, & Winston, 1985), 66–67.

44. Garson, *Electronic Sweatshop,* 128–154.

45. Robert Blauner, *Alienation & Freedom: The Factory Worker and His Industry* (Chicago: University of Chicago Press, 1964).

46. See William A. Faunce, "Automation and the Division of Labor," *Social Problems* 13, no. 2 (1965): 149–60 and Martin Meissner, *Technology and the Worker: Technical Demands and Social Processes in Industry* (San Francisco: Chandler Publishing Co., 1969).

47. Paul S. Adler, "New Technologies, New Skills," *California Management Review* 29, no. 1 (1986): 9–28. For similar upgrading results in a study of two U.S. banks employing computerized technology two decades earlier, see Edward R.F.W. Crossman, Stephen Laner, Louis E. Davis, and Stanley H. Caplan, *Evaluation of Changes in Skill-Profile and Job-Content Due to Technological Change: Methodology and Pilot Results from the Banking, Steel and Aerospace Industries,* Report to the Office of Manpower Policy, Evaluation and Research, U.S. Department of Labor Contract 81–04–05 (Berkeley: Department of Industrial Engineering and Operations Research, University of California, Berkeley, 1966).

48. See Paul Attewell, "Skill and Occupational Changes in U.S. Manufacturing," in *Technology and the Future of Work,* ed. Paul S. Adler (New York: Oxford University Press, 1992), 46–88.

49. This example also reveals problems with another claim by Levy and Murnane, namely that the same occupations that are at risk of computer substitution are also at risk of offshoring. They make this claim on the grounds that the type of work that gets offshored can, like the work prone to computer substitution, be explained in clear rules. However, the radiologists in places like India carried out their tasks in a manner nearly identical to their U.S. counterparts in that they did not follow rule-based logic when making their diagnoses, but recognized patterns.

50. By 2005, Levy and his colleagues well understood the role of economic, professional, and regulatory factors in shaping computer use in the work of diagnostic radiologists. In fact, Levy and colleague Ari Goelman considered exactly these factors in the specific context of diagnostic radiology in a 2005 article in the Brookings Trade Forum. Yet, they failed to acknowledge the problems that the case of radiology poses for Levy and Murnane's formulation, which they presented as a sort of

general rule that can explain many situations of computerization or offshoring of work. See Frank Levy and Ari Goelman, "Offshoring and Radiology," *Brookings Trade Forum*, 2005, 411–423.

51. See L. Jarvis and B. Stanberry, "Teleradiology: Threat or Opportunity?," *Clinical Radiology* 60, no. 8 (2005): 840–845. See also Martin Stack, Myles Gartland, and Timothy Keane, "The Offshoring of Radiology: Myths and Realities," *SAM Advanced Management Journal* 72, no.1 (2007): 44–51.

52. A 2007 survey of 405 American College of Radiology (ACR) members found that 45 percent of U.S. radiologists used teleradiology services, up from 15 percent of a larger sample (1321 respondents) in 2003. In a 2011 survey of 363 radiology groups across the country, CapSite, a healthcare technology research and advisory firm, found a much higher value, with 63 percent of the groups they surveyed employing teleradiology services. The 2007 ACR survey found no significant differences in the use of teleradiology services by geographic region of the United States, but large groups (30 or more radiologists) were less likely than medium-sized (15–29 radiologists) and small groups (1–10 radiologists) to use teleradiology services, and nonmetropolitan groups had low odds of doing so. In an accompanying study, the authors speculated that teleradiology was a key factor in explaining how radiologists achieved a productivity increase nearly twice that of most workers in the U.S. economy over the period from 1992 to 2007. See Rebecca S. Lewis, Jonathan H. Sunshine, and Mythreyi Bhargavan, "Radiology Practices' Use of External Off-Hours Teleradiology Services in 2007 and Changes since 2003," *American Journal of Roentgenology* 193, no. 5 (2009): 1333–1339; and Mythreyi Bhargavan, Adam H. Kaye, Howard P. Forman, and Jonathan H. Sunshine, "Workload of Radiologists in United States in 2006-2007 and Trends since 1991–1992," *Radiology* 252, no. 2 (2009): 458–467. For an example of the profession's activities in shaping the use of teleradiology services, see Arl Van Moore et al., "Report of the ACR Task Force on International Teleradiology," *Journal of American College of Radiology* 2, no. 2 (2005): 121–125.

53. See Max Weber, *Economy and Society*, edited by Guenther Roth and Claus Wittich (New York: Bedminister Press, 1968), vol. 1, *Conceptual Exposition,* 342–345. See also John Van Maanen and Stephen R. Barley, "Occupational Communities: Culture and Control in Organizations," in *Research in Organizational Behavior* 6, ed. Barry M. Staw and Larry L. Cummings (Greenwich, CT: JAI Press, 1984), 287–365.

54. See Stephen R. Barley and Julian E. Orr, "Introduction: The Neglected Workforce," in *Between Craft and Science: Technical Work in U.S. Settings*, ed. Stephen R. Barley and Julian E. Orr (Ithaca, NY: ILR Press, 1997), 1–19.

55. See, for example, Gary Gereffi, Vivek Wadhwa, Ben Rissing, and Ryan Ong, "Getting the Numbers Right: International Engineering Education in the United States, China, and India," *Journal of Engineering Education* 97, no. 1 (2008): 13–25; National Academy of Engineering, Committee on Science, Engineering, and Public Policy, *Rising Above the Gathering Storm: Energizing and Employing America for a Brighter*

Economic Future (Washington, DC: National Academy Press, 2006); Geoffrey Colvin, "America Isn't Ready [Here's What to Do About it]," *Fortune International* (Europe) 152, no. 3 (August 8, 2005): 22–31; and H. Oner Yurtseven, "How Does the Image of Engineering Affect Student Recruitment and Retention? A Perspective from the USA," *Global Journal of Engineering Education* 6, no. 1 (2002): 17–23.

56. See, for example, Elizabeth Jessup, Tamara Sumner, and Lecia Barker, "Report from the Trenches: Implementing Curriculum to Promote the Participation of Women in Computer Science," *Journal of Women and Minorities in Science and Engineering* 11, no. 3 (2005): 273–294; and Michelle J. Johnson and Sheri D. Sheppard, "Relationships between Engineering Student and Faculty Demographics and Stakeholders Working to Affect Change," *Journal of Engineering Education* 93, no. 2 (2004): 139–151.

Chapter 2

1. See Louis L. Bucciarelli, *Designing Engineers* (Cambridge, MA: MIT Press, 1994); and Kathryn Henderson, *On Line and On Paper: Visual Representations, Visual Culture, and Computer Graphics in Design Engineering* (Cambridge, MA: MIT Press, 1999).

2. For FEA analyses, IAC, like most automakers, used either NASTRAN (which MSC Software developed for NASA) or DYNA3D (which Lawrence Livermore National Laboratories developed) solvers; for CFD analyses, it used Abaqus (a solver that Dassault developed).

3. A plenum is a ventilation duct designed to allow air circulation for heating and air conditioning systems. To avoid obstructing air flow, no parts should penetrate the plenum.

4. A loadcase was a particular set of boundary conditions, often specified by a government agency as a standard safety test. An example of a loadcase was the set of conditions representing a 35 mile-per-hour frontal offset deformable barrier test.

5. Our field notes of this event were longer and more detailed than this segment suggests. We have trimmed the notes to include only those details necessary to understand the event, a practice we maintain through the book.

6. D stands for "dead load," or the load of the building itself, primarily from its components; L stands for "live load," or the load from the building's occupants and their furnishings; E stands for "earthquake load" and refers to the added load that an earthquake engenders.

7. W10 × 33 denotes a wide flange beam that is 10 inches deep and 33 pounds per linear foot.

8. A kip, like a Newton, is a unit of force; 1 kip = 1,000 pounds of force.

9. Diag is short for diagnostic, another name for test. Literals are constant values, as opposed to variables, whose values can change.

10. One of the paradoxes of chip design was that "software" engineers worked at desks and in labs covered with physical hardware in the form of circuit boards, small motors, fans, cables, zip drives, and network controllers, while "hardware" engineers sat in front of computer monitors typing code at desks barren of computer paraphernalia save for a design manual or two.

Chapter 3

1. The writer Mario Salvadori echoed Sam's sentiment when he noted, "No structural engineer accepts the output of a computer run unless it agrees (more or less) with what experience tells him to be the correct answer." See Mario Salvadori, *Why Buildings Stand Up: The Strength of Architecture* (New York: Norton, 1980), 21.

2. Sam's account is evidence of how structural engineers think visually, a point that Eugene S. Ferguson makes eloquently and in detail in his book, *Engineering and the Mind's Eye* (Cambridge, MA: MIT Press, 1994).

3. In chip design, a netlist is a list of logic gates and their interconnections. In other electronics design domains, the term has slightly different but largely similar meanings.

4. For a thorough description and explanation of how people gain sentient knowledge from physical objects at work, and how shifting to computerized work affects people's subsequent ways of knowing and thinking, see Shoshana Zuboff, *In the Age of the Smart Machine: The Future of Work and Power* (New York: Basic Books, 1988).

5. Engineers called the barrier "offset" because it was less than the width of the vehicle and not centered; the engineers placed the barrier so that only the right side or the left side of the front of the vehicle would strike the barrier head-on, while the other side would remain more or less unscathed. They called the barrier deformable because the vehicle would not take the full force of the impact; the barrier would also absorb some of that force.

Chapter 4

1. See, for example, James D. Thompson, *Organizations in Action: Social Science Bases of Administrative Theory* (New York: McGraw-Hill, 1967); Gregory P. Shea and Richard A. Guzzo, "Groups as Human Resources," in *Research in Personnel and Human Resources Management*, ed. Gerald R. Ferris and Kendrith M. Rowlands, 323–367 (Greenwich, CT: JAI Press, 1987); Ruth Wageman, "Interdependence and Group Effectiveness," Administrative Science Quarterly 40, no.1 (1995): 145–180; and Daniel G. Bachrach, Benjamin C. Powell, Elliot Bendoly, and R. Glenn Richev,

"Organizational Citizenship Behavior and Performance Evaluations: Exploring the Impact of Task Interdependence," *Journal of Applied Psychology* 91, no. 1 (2006): 193–201.

2. See, for example, Ann E. Gray, Abraham Seidmann, and Kathryn E. Stecke, "A Synthesis of Decision Models for Tool Management in Automated Manufacturing," *Management Science* 39, no. 5 (1993): 549–567.

3. We draw on our previous paper for these concepts and some of the findings that we discuss in this chapter. See Diane E. Bailey, Paul M. Leonardi, and Jan Chong, "Minding the Gaps: Understanding Technology Interdependence and Coordination in Knowledge Work," *Organization Science* 21, no. 3 (2010): 713–730. That paper compared only structural and hardware engineers; in this chapter, we expand our analysis to encompass all three occupations.

4. See Gerben S. Van der Vegt and Evert Van de Vliert, "Effects of Perceived Skill Dissimilarity and Task Interdependence on Helping in Work Teams," *Journal of Management* 31, no. 1 (2005): 73–89, esp. 78.

5. A kip is a kilo-pound; one kip equals 1,000 pounds.

6. See Rich Haswell, "Automated Text-checkers: A Chronology and a Bibliography of Commentary," Department of English, Bowling Green State University, 2005, http://www.bgsu.edu/departments/english/cconline/haswell/haswell.htm.

7. In Thompson's terms, technology interdependence in all three occupations was largely *sequential*, with each technology depending on the previous one for input and providing to the next one output. *Reciprocal* technology interdependence arose in the case of rework, or when a technology took as input its follower's output. In automotive engineering, task interdependence was primarily sequential, with each engineer relying on the previous one for input and providing to the next one output. But in structural and hardware engineering, task interdependence was *pooled*: the engineers worked on distinct components, which they combined at the end, with no engineer immediately dependent on another for input. We say "immediately" because some interdependence did exist. For example, each hardware engineer's code "called" the code of many of his peers, but the engineer did not need his peers' code to write his own. Rather, trusting input specifications, he could assume his peers' code would take input and output in a certain way, and then write his own code based on these assumptions. In this manner, input specifications limited task interdependence. A similar situation existed in structural engineering. Although the physics of building structures compelled each floor to bear the load of all the floors above it, a fact that might suggest the sequential design of a building from its uppermost floor to its lowermost, in practice, engineers used estimates of other floor loads when designing their own floor. In both hardware and structural engineering, engineers made informed assumptions about inputs to limit their task interdependence. See Thompson, *Organizations in Action*.

8. Vera is a hardware verification language that engineers used to write tests of RTL models that they had created in hardware description languages such as Verilog and VDHL.

9. See www.edacafe.com and www.deepchip.com.

10. Mario Salvadori, *Why Buildings Stand Up: The Strength of Architecture* (New York: Norton, 1980), 73.

11. The "ASD manual" refers to the *Manual of Steel Construction: Allowable Stress Design*, a publication of the American Institute of Steel Construction.

12. We draw here on Paul's case study of this group; see Paul M. Leonardi, *Car Crashes without Cars: Lessons about Simulation Technology and Organizational Change from Automotive Design* (Cambridge, MA: MIT Press, 2012).

13. Frederick Winslow Taylor has long drawn the attention of management historians and organizational scholars. His 1913 *Principles of Scientific Management* (Mineola, NY: Dover Publications, 1998) is a management classic; see Arthur G. Bedeian and Daniel A. Wren, "Most Influential Management Books of the 20th Century," *Organizational Dynamics* 29, no. 3 (2001): 221–225. Peter F. Drucker, himself a prominent management writer, considered Taylor to be the twentieth century's foremost thinker on the topics of productivity, work, and management; see Peter F. Drucker, "Knowledge-Worker Productivity: The Biggest Challenge," *California Management Review* 41, no. 2 (1999): 79–94. For a good history of Taylor and the movement that arose around his management principles, see Daniel Nelson, *Frederick W. Taylor and the Rise of Scientific Management* (Madison: University of Wisconsin Press, 1980).

14. For Ford's denial of a Tayloristic influence, see David A. Hounshell, *From the American System to Mass Production, 1800–1932* (Baltimore, MD: Johns Hopkins University Press, 1984), 249–253.

Chapter 5

1. For a complete discussion of how engineers taught and learned from one another in the course of everyday work, including the teaching methods we describe in this chapter and the frequencies of their use, see Diane E. Bailey and Stephen R. Barley, "Teaching-Learning Ecologies: Mapping the Environment to Structure through Action," *Organization Science* 22, no. 1 (2011): 262–285.

2. RFI stands for "Request for Information." Contractors submitted RFIs to the structural engineers whenever the structural engineering drawings were unclear or incomplete, or whenever the contractors were otherwise uncertain how to proceed in the field.

3. Unistrut was a manufacturer of metal framing systems used to support industrial catwalks, piping systems, solar panels, and the like; the plural noun here referred to framing elements.

4. TS2×2 was a beam specification in which TS stood for tube steel. Tube steel beams were hollow beams with square or rectangular external shapes, in contrast to W beams, or wide flange beams, which were shaped like an "I" and which previously engineers colloquially called "I-beams." The 2×2 notation referred to 2 inches deep by 2 pounds per linear foot.

5. For details on which countries send and receive offshored software programming work, see William Aspray, Frank Mayadas, and Moshe Y. Vardi eds., *Globalization and Offshoring of Software, A Report of the ACM Job Migration Task Force* (New York: Association for Computing Machinery, 2006), 52–54.

6. See Cristina Chaminade and Jan Vang. "Globalisation of Knowledge Production and Regional Innovation Policy: Supporting Specialized Hubs in the Bangalore Software Industry," *Research Policy* 37, no. 10 (2008): 1684–1696.

7. For an introduction to this literature, see Pamela Hinds and Sara Kiesler, eds., *Distributed Work* (Cambridge, MA: MIT Press, 2002).

8. ODB stands for offset deformable barrier.

9. The "body-in-white" refers to the welded sheet metal components that form the vehicle's structure and is the load-bearing frame onto which the vehicle's other components, such as the engine, transmission, and exterior and interior trim, will be affixed.

10. For a complete discussion of how and why some engineers at global centers did take the time to teach their India center counterparts and others did not, see Paul M. Leonardi and Diane E. Bailey, "Transformational Technologies and the Creation of New Work Practices: Making Implicit Knowledge Explicit in Task-Based Offshoring," *MIS Quarterly* 32, no. 2 (2008): 411–436.

Chapter 6

1. See Anselm Strauss and Juliet Corbin, *Basics of Qualitative Research: Techniques and Procedures for Developing Grounded Theory,* 2nd ed. (Thousand Oaks, CA: Sage, 1998).

2. For a more in-depth treatment of the implicit knowledge in artifacts and how engineers went about trying to make it explicit, see our earlier paper: Paul M. Leonardi and Diane E. Bailey, "Transformational Technologies and the Creation of New Work Practices: Making Implicit Knowledge Explicit in Task-Based Offshoring," *MIS Quarterly* 32, no. 2 (2008): 411–436.

Chapter 7

1. Paul M. Leonardi and Stephen R. Barley, "What's under Construction Here? Social Action, Materiality, and Power in Constructivist Studies of Technology and Organizing," *Academy of Management Annals* 4 (2010): 1–51.

2. See, for example, Howard S. Becker and James Carper, "The Elements of Identification with an Occupation," *American Sociological Review* 21 no. 3 (1956): 341–348. See also Everett C. Hughes, *Men and Their Work* (Glencoe, IL: Free Press, 1958), Andrew Abbott, *The System of Professions: An Essay on the Division of Expert Labor* (Chicago: University of Chicago Press, 1988), and Stephen R. Barley and Gideon Kunda, "Bringing Work Back In," *Organization Science* 12, no. 1 (2001): 76–95.

3. See, for example, W. Richard Scott. *Institutions and Organizations: Ideas and Interests* (Thousand Oaks, CA: Sage, 2007). See also Paul J. DiMaggio and Walter W. Powell, "The Iron Cage Revisited: Institutional Isomorphism and Collective Rationality in Organizational Fields," *American Sociological Review* 48 (1983): 147–160, and Kevin T. Leicht and Mary L. Fennell, "Institutionalism and the Professions," in *The Sage Handbook of Organizational Institutionalism,* ed. Royston Greenwood, Christine Olver, Kerstin Sahlin, and Roy Suddaby, 431–438 (Thousand Oaks, CA: Sage, 2008).

4. See, for example, work by Martin Fischer, John Haymaker, and Kathleen Liston, "Benefits of 3D and 4DModels for Facility Managers and AEC Service Providers," in *4D CAD and Visualization in Construction: Developments and Applications*, ed. Raja R. A. Issa, Ian Flood, and William J. O'Brien, 1–32 (Lisse, The Netherlands: A. A. Balkema, 2003). See also K. W. Chau, M. Anson, and J. P. Zhang, "4D Dynamic Construction Management and Visualization Software: 1. Development," *Automation in Construction* 14, no. 4 (2005): 512–524. Adoption of visualization software in construction firms can also be problematic for a variety of organizational, business, and human reasons (not necessarily technical ones). See, for example, Jennifer Whyte, Dino Bouchlaghem, and Tony Thorpe, "IT Implementation in the Construction Organization," *Engineering, Construction, and Architectural Management* 9, no. 5/6 (2002): 371–377.

5. For an example of an early conceptualization of knowledge management systems in structural engineering, see J. M. Kamara, G. Augenbroe, C. J. Anumba, and P. M. Carrillo, "Knowledge Management in the Architecture, Engineering and Construction Industry," *Construction Innovation: Information, Process, Management* 2, no. 1 (2002): 53–67.

6. See Levin Institute, *The Evolving Global Talent Pool: Issues, Challenges, and Strategic Implications* (New York: Levin Institute, 2005), www.levin.suny.edu/pdf/globalTalentPool.pdf.

7. Many writers in government, education, industry, and media have engaged in this debate over numbers in the past decade. A good place to start to understand the

issues involved is with a paper by globalization and engineering scholars at Duke University. See Gary Gereffi, Vivek Wadhwa, Ben Rissing, and Ryan Ong, "Getting the Numbers Right: International Engineering Education in the United States, China, and India," *Journal of Engineering Education* 97, no. 1 (2008): 13–25.

8. Paul Otellini, "How the Private Sector Can Curb Our Engineering Shortage," *Washington Post,* August 4, 2011.

9. See William P. Butz, Gabrielle A. Bloom, Mihal E. Gross, Terrence K. Kelly, Aaron Kofner, and Helga E. Rippen. "Is There a Shortage of Scientists and Engineers? How Would We Know?" (Santa Monica, CA: RAND Corp., 2003). In addition to providing policy recommendations and cautioning against policy missteps, this paper shows the various ways in which policy makers might conceptualize what a shortage of scientists and engineers is, and how to measure shortage under various conceptualizations.

10. Gereffi et al., "Getting the Numbers Right."

11. See Diana Farrell, Martha Laboissière, Jaeson Rosenfeld, Sascha Stürze, and Fusayo Umezawa, *The Emerging Global Labor Market: Part II. The Supply of Offshore Talent in Services* (San Francisco, CA: McKinsey Global Institute, 2005).

12. See Stephan Manning, Silvia Massini, and Arie Y. Lewin, "A Dynamic Perspective on Next-Generation Offshoring: The Global Sourcing of Science and Engineering Talent," *Academy of Management Perspectives* 22, no. 3 (2008): 35–54.

13. See Ashok Deo Bardhan and Cynthia A. Kroll, *The New Wave of Outsourcing,* Fisher Center Research Report 1103 (Berkeley: University of California, 2003).

14. See Alan S. Blinder, "How Many US Jobs Might Be Offshorable?," *World Economics* 10, no. 2 (2009): 41–78.

15. Erik Brynjolfsson and Andrew McAfee, *The Second Machine Age: Work, Progress, and Prosperity in a Time of Brilliant Technologies* (New York: W.W. Norton & Co., Inc., 2014), 256.

16. Lori Rosnkopft and Michael Tushman make a similar point when they argue that technological determinism and social constructivism are not in conflict but proceed through alternating periods in the lifecycle of a technology. For more details, see Lori Rosenkopf and Michael L. Tushman, "The Coevolution of Technology and Organization," in *Evolutionary Dynamics of Organizations,* ed. J. A. C. Baum and J. Singh, 403–424 (New York: Oxford University Press, 1994).

17. Marie-Claude Boudreau and Daniel Robey, "Enacting Integrated Information Technology: A Human Agency Perspective," *Organization Science* 16, no. 1 (2005): 3–18.

18. See Marcie J. Tyre and Wanda J. Orlikowski, "Windows of Opportunity: Temporal Patterns of Technological Adaptation in Organizations," *Organization Science* 5,

no. 1 (1994): 98–118. See also Wanda J. Orlikowski, "The Duality of Technology: Rethinking the Concept of Technology in Organizations," *Organization Science* 3, no. 3 (1992): 398–426.

19. See Mark Dodgson, David M. Gann, and Ammon Salter, "'In Case of Fire, Please Use the Elevator': Simulation Technology and Organization in Fire Engineering," *Organization Science* 18, no. 5 (2007): 849–864. Later in this section, we quote passages from pages 856 and 860, respectively.

Appendix

1. We did study, though briefly, one more site. One of the hardware engineering firms in our study had recently acquired an analog design firm with a design center in Austin, Texas. Diane and two students spent a single week in Austin observing these engineers. Our main reason for going to that site was that the host firm was quite insistent we do so, being very excited about their recent acquisition and eager for us to see the differences between analog design and digital design, which is what our three hardware engineering firms otherwise practiced. To gain insights into these design differences, we included this firm's Austin site in our study. However, because analog design employed a different technology suite from that used in digital design, and because we had no comparison group of analog designers to observe beyond the Austin site, we did not expand our fieldwork there and excluded the Austin data we did have from the discussion and analyses of this book.

2. See Paul M. Leonardi, *Car Crashes Without Cars: Lessons about Simulation Technology and Organizational Change from Automotive Design* (Cambridge, MA: MIT Press, 2012).

3. See Gary L. Downey, Arthur Donovan, and Timothy J. Elliott, "The Invisible Engineer: How Engineering Ceased to Be a Problem in Science and Technology Studies," in *Knowledge and Society: Studies in the Sociology of Science Past and Present,* ed. Lowell L. Hargens, Robert A. Jones, and Andrew Pickering (Greenwich, CT: JAI, 1989), 189–216. See also Stephen R. Barley, "What We Know (and Mostly Don't Know) about Technical Work," in *The Oxford Handbook of Work and Organization,* ed. Stephen Ackroyd, Rosemary Batt, Paul Thompson, and Pamela S. Tolbert (Oxford: Oxford University Press, 2005), 376–403.

4. For an ethnography of a street, see Mitchell Duneier, *Sidewalk* (New York: Farrar, Straus and Giroux, 1999). For an ethnographer's account of gaining entrée to a street gang, see Sudhir Venkatesh, *Gang Leader for a Day: A Rogue Sociologist Takes to the Streets* (New York: Penguin Press, 2008). For a description of how an ethnographer worked in a factory, see Laurie Graham, *On the Line at Subaru-Isuzu: The Japanese Model and the American Worker* (Ithaca, NY: ILR Press, 1995). To read about legitimate peripheral participation in communities of practice, see Jean Lave and Etienne

Wenger, *Situated Learning: Legitimate Peripheral Participation* (New York: Cambridge University Press, 1991).

5. Ethnographers often talk of the roles they played in the field as a means of demonstrating to readers how they gained acceptance into the scene. Charles Bosk, for example, in his ethnography of surgeons in training, wrote how he was, among other roles, gofer, referee, historian, advisee, and clown. See Charles L. Bosk, *Forgive and Remember: Managing Medical Failure* (Chicago: University of Chicago Press, 1979). Our role was only rarely more than "researcher." Occasionally, we were "student." The role of student, perhaps naturally, fell most comfortably on the students on our team. Senior engineers in particular were fond of taking our student researchers off for a quick display of a new technology. At times Diane or Paul served as "adviser" to senior engineers who, during a chance meeting or as part of an interview, hoped to glean some insights into their firm's work practices. For example, at one structural engineering firm, when Diane noted in conversation with a principal of the firm the mentoring the team had observed, he expressed surprise because a junior person had just quit on grounds of insufficient attention. That led to a discussion of how our observations revealed that mentoring was never a formal "Let's chat Thursday at three" event, and how junior engineers were perhaps so caught up in defending their work that they failed to recognize learning opportunities that arose in the course of design reviews. On several occasions, Paul gave managers at India's IAC site debriefs that focused on a handful of take-aways that helped them understand their situation vis-à-vis the rest of the IAC engineering workforce, which we visited more than they did. For Diane and, later, Paul, as professors who could not easily adopt the role of student, we had to work a bit harder than our student researchers to get informants comfortable with us. We joked with our informants to the extent that the culture and the workload permitted; hardware engineers, for example, were keen to know how they differed from structural engineers and were amused when we told them that, unlike hardware engineers, structural engineers did not think free food was part of their benefits package and never used "build" as a noun. We interviewed key informants over lunch on occasion, which we gave us a chance to talk comfortably outside the workplace.

6. For a discussion of the importance of such close reading of the data, see Robert M. Emerson, Rachel I. Fretz, and Linda L. Shaw, *Writing Ethnographic Fieldnotes* (Chicago: University of Chicago Press, 1995). See also Anselm Strauss and Juliet Corbin, *Basics of Qualitative Research: Techniques and Procedures for Developing Grounded Theory*, 2nd ed. (Thousand Oaks, CA: Sage, 1998).

7. Robert K. Yin, *Case Study Research: Design and Methods* (Beverly Hills, CA: Sage, 1994).

8. Matthew B. Miles and A. Michael Huberman, *Qualitative Data Analysis: A Source Book of New Methods*, 2nd ed. (Beverly Hills, CA: Sage, 1994).

9. See Diane E. Bailey, Paul M. Leonardi, and Jan Chong, "Minding the Gaps: Understanding Technology Interdependence and Coordination in Knowledge Work," *Organization Science* 21, no. 3 (2010): 713–730.

10. Barney Glaser, *Theoretical Sensitivity* (Mill Valley, CA: Sociological Press, 1978).

11. See Diane E. Bailey and Stephen R. Barley, "Teaching-Learning Ecologies: Mapping the Environment to Structure through Action," *Organization Science* 22, no. 1 (2011): 262–285.

12. Barker detailed his research and methods in numerous publications. To begin an exploration of his work, see Roger G. Barker, *The Stream of Behavior* (New York: Appleton-Century-Crofts, 1963).

13. See Philip Schoggen, *Behavior Settings: A Revision and Extension of Roger G. Barker's Ecological Psychology* (Stanford, CA: Stanford University Press, 1989).

14. See Diane E. Bailey, Paul M. Leonardi, and Stephen R. Barley, "The Lure of the Virtual," *Organization Science* 23, no. 5 (2012): 1485–1504.

15. See Jack Katz, "From How to Why: On Luminous Description and Causal Inference in Ethnography (Part 2)," *Ethnography* 2, no. 4 (2002): 443–473.

16. In qualitative research, triangulation entails the use of multiple data sources or methods to confirm a conclusion drawn from the data. For a discussion of triangulation, see Miles and Huberman, *Qualitative Data Analysis,* 2nd ed., 266–267.

17. We appreciate the irony that we consider more than a decade in the field to constitute fast movement on our part. It was.

Index